Maths for Computing and Information Technology

Essential Maths for Students

Series editors: Anthony Croft and Robert Davison
Department of Mathematical Sciences,
De Montfort University, Leicester.

Also available

Foundation Maths
Croft and Davison

Maths and Statistics for Business
Lawson, Hubbard and Pugh

Forthcoming

Maths and Statistics for the Built Environment
Bacon

Essential Maths for Students

Maths for Computing and Information Technology

Frank Giannasi
and
Robert Low

Longman Scientific & Technical
Longman Group Limited
Longman House, Burnt Mill, Harlow
Essex CM20 2JE, England
and Associated Companies throughout the world

First published 1995

British Library Cataloguing in Publication Data
A catalogue entry for this title is available from the British Library.

ISBN 0-582-23654-1

Typeset by INW in 10/12pt Times New Roman

Produced by Longman Singapore Publishers (Pte) Ltd.
Printed in Singapore

Typeset by 1WW in 10/12pt Times New Roman

Produced by Longman Singapore Publishers (Pte) Ltd.
Printed in Singapore

Contents

Preface

Computing and Information Technology are both subjects which require a certain degree of mathematical knowledge on the part of their practitioners. However, the widening of access to higher education has enabled students without the traditional mathematical training to undertake courses in these areas, and even those students who have studied mathematics at school will find that many of the relevant topics are not part of the syllabus they have studied.

Maths for Computing and Information Technology is intended to cover those aspects of mathematics useful for students studying computing and related subjects. It is intended for those who require a basic grounding in the relevant mathematics, and assumes only a knowledge of basic algebra.

Throughout, the style of the book is to introduce ideas and techniques through the medium of examples. Each chapter begins with a statement of the objectives of that chapter, detailing what abilities a successful study of the chapter should impart. Most sections conclude with self-assessment questions, which provide a quick guide to how well the material has been understood. The self-assessment questions are followed by a set of exercises on the material of the section, to which complete answers are provided at the end of the book. Each chapter is followed by a set of test exercises from which a tutor can set assignments or tests; answers to these are not provided.

We recommend the following strategy for using the book. Treat the sections as units. To work through a section, read it carefully, paying particular attention to the worked examples. Use the self-assessment questions to indicate whether you understand the material. When you have understood the examples, and before attempting the exercises, try to do the examples without reference to the solutions. Next, attempt the exercises at the end of the section, checking your answers with those provided at the back of the book.

Remember, mastering a piece of mathematics requires application; doing the exercises builds up your ability to apply the ideas to different problems, but requires a certain amount of effort. As you practise, though, you should find that the work becomes easier and the new ways of thinking become more natural.

We hope you find this text useful, and wish you good luck in your studies.

Frank Giannasi and Robert Low

1 Propositional calculus

Objectives

When you have mastered this chapter you should be able to

- translate between natural language and symbolic notation
- use tabular methods to reason about the consistency of collections of statements and the validity of arguments
- use algebraic methods to demonstrate the equivalence of statements
- construct and read proofs using rules of inference

In this chapter, we will meet part of the language of mathematics; in particular, we will get to grips with some of the mathematics used for reasoning and constructing proofs. This part of mathematics is called propositional calculus, and is of particular relevance in the area of reasoning about and deriving programs. It is also important in the area of specification, as it allows one to reason about the consistency of a specification. Finally, the propositional calculus provides a formidable tool for thought, as it gives a framework for analysing arguments in general.

1.1 Propositions

In natural English, a **proposition** is a statement, *i.e.* a sentence which may be either true or false. For example,

1. Two plus two is four.
2. E. H. Shepard wrote *Winnie the Pooh*.
3. King Henry the Eighth had six wives.

are propositions, the first and third of which are true, the second of which is false.

In contrast,

1. It's too hot.
2. John is tall.

are not propositions, as they cannot unambiguously be described as true or false.

Worked example

1.1 Which of the following are propositions?

1. Squares have four sides.
2. Soap operas are fun to watch.
3. If I take my umbrella, it's bound to be sunny.
4. How are you today?
5. Triangles have four sides.

Solution The first, third and fifth are propositions.

A proposition may be described as **simple** if it cannot be broken up into sub-propositions, and **compound** if it can.

Example

1.2 The following are simple propositions:

1. My program compiles.
2. My program has no run-time errors.
3. My program is correct.

whereas the following are compound propositions:

1. My program has no run-time errors, and my program is correct.
2. If my program is correct, then my program compiles.

Self-Assessment Questions 1.1

1. What is meant by a proposition?
2. Why is 'What are you doing tonight?' not a proposition?
3. What kind of proposition is 'If this is Saturday, then I have no lectures'?

Exercise 1.1

Classify the following sentences as simple propositions, compound propositions or not propositions.

1. If my program is correct, it will pass the validation suite.
2. Compile that program now!
3. If my program passes the validation suite, it is correct.
4. Mars is inhabited by little green men.
5. Hacking is good fun.

1.2 Symbolic logic

If we want to be able to carry out calculations about propositions, *i.e.* to have a way of analysing statements and arguments, we will find it convenient to have a concise notation. We will use one of the standard notations for propositional calculus, though there are others.

We will approach this symbolic technique in two steps; first we will see how symbols may be used to represent ways of connecting together propositions, then we will see how the use of letters to represent (simple) propositions may be used to make everything more concise.

Logical connectives

First, we consider the standard **logical connectives**. These are ways of combining propositions together; any proposition may be constructed out of simple propositions using the connectives we consider. Note that the set of connectives we will use is not the only one possible, though it is a particularly popular one.

Conjunction

The two propositions 'My Pascal compiler only accepts ANSI standard Pascal' and 'My program compiles under my compiler' may be combined into the proposition 'My Pascal compiler only accepts ANSI standard Pascal, *and* my program compiles under my Pascal compiler'. This new proposition is called the **conjunction** of the two original propositions, and is true precisely when they are both true.

Worked example

1.3 What is the conjunction of the following two propositions, and when is the conjunction true?

1. The system specification is acceptable.
2. The programming team will start development today.

Solution 'The system specification is acceptable *and* the programming team will start development today.' This proposition is true when both parts are true, and only then.

The symbol used to represent conjunction is $\wedge$; a special symbol is used to remind us that conjunction is what we mean and that the everyday English meanings of the word 'and' are not relevant. Using this symbol, then, the propositions above are written 'My Pascal compiler only accepts ANSI standard Pascal $\wedge$ my program compiles under my Pascal compiler' and 'The system specification is acceptable $\wedge$ the programming team will start development today'.

Disjunction

This is the first logical connective that is liable to cause problems. It is the connective that corresponds to the English 'or', but means something slightly different. **Disjunction** is denoted by the symbol $\vee$, and the distinction between English 'or' and logical $\vee$ may exemplified as follows. If we consider the two propositions

1. You can have chips with this meal.
2. You can have mashed potatoes with this meal.

then the English sentence 'You can have chips with this meal or you can have mashed potatoes with this meal' is generally understood as meaning that you can have whichever one of the two you want, but you cannot have both. The logical connective, however, has a somewhat different meaning. 'You can have chips with this meal $\vee$ you can have mashed potatoes with this meal' means that at least one of the statements 'You can have chips with this meal' and 'You can have mashed potatoes with this meal' is true. So, a waitress using 'or' in its logical sense might say 'You can have chips *or* you can have mashed potato', then, when you asked for chips, tell you that only mashed potato was available. For this reason, you must be careful how you convert English descriptions into symbolic logic form.

Worked example

1.4 What is the disjunction of the statements

1. The specification is adequate.
2. The programming team is happy with the specification.

Is the statement 'The specification is not adequate' compatible with the disjunction?

Solution The disjunction is 'The specification is adequate $\vee$ the programming team is happy with the specification'. The given statement is compatible with this disjunction if at least one of the parts is true, so as long as the programming team is happy with the specification, the disjunction is true.

Implication A fundamental part of any argument is **implication**. Again, the logical use is rather different from the everyday use of implication. In normal conversation, when we say that one statement implies another, we usually have in mind some way in which the truth of the first statement forces that of the second. Logical implication simply says that whenever the first statement is true, the second is also; the standard terminology for the first statement is the **premiss**, or alternatively the **antecedent**, and for the second it is the **consequent**. The symbol $\Rightarrow$ is used to denote this relationship. Thus, the proposition 'The specification is suitable $\Rightarrow$ the programming team is happy' simply tells us that whenever the specification is suitable, the programming team is happy; it does not say that the suitability of the specification causes the happiness of the programming team.

Worked example

1.5 If the proposition 'the specification is suitable $\Rightarrow$ the programming team is happy' is true, could it be the case that the specification was unsuitable but the programming team was, nevertheless, happy?

Solution Yes, since the implication only tells us that whenever the specification is suitable the programming team is happy; it says nothing about the case when the specification is unsuitable.

There are further issues worthy of discussion here, but we will postpone these until we have a more concise notation.

Logical equivalence The **truth value** of a proposition is its truth or falsity; a proposition that is false has truth value F, one that is true has truth value T. Two propositions are said to be *logically equivalent* if they always have the same truth value. Logical equivalence is denoted by the symbol $\Leftrightarrow$.

Example

1.6 The proposition 'The specification is adequate $\Leftrightarrow$ the programming team is happy' means that either the specification is adequate and the programming team is happy or the specification is not adequate and the programming team is not happy.

Negation Strictly speaking, **negation** is not a connective, but this is an appropriate place to introduce it anyway. The negation of a proposition is the proposition that is true whenever the original proposition is false, and *vice versa*. We denote the negation of a proposition by prefixing the symbol $\neg$ to it.

Worked example

1.7 What is the negation of the proposition 'The programming team is happy'?

Solution In normal English, 'The programming team is not happy', or, using the symbolic notation, '¬the programming team is happy'.

Making the notation concise

At this point, it is fairly clear that what we have been doing could easily get out of hand if the simple propositions used in an argument, specification or whatever, are long English sentences. For this reason, it is customary to use single letters (or words) to represent simple propositions. It is also a good idea (though not quite as customary) to use letters or words that have some mnemonic value relating them to the propositions they represent.

Worked example

1.8 Using the letter S to stand for the proposition 'The specification is suitable' and the letter H to stand for 'The programming team is happy', re-express

1. The specification is suitable and the programming team is happy.
2. If the specification is suitable, then the programming team is happy.
3. The specification is not suitable, and the programming team is not happy.
4. If the programming team is happy, then the specification is suitable.
5. The specification is suitable precisely when the programming team is unhappy.

Solution

1. $S \wedge H$
2. $S \Rightarrow H$
3. $(\neg S) \wedge (\neg H)$ (Note that we use brackets to indicate that we are expressing the conjunction of two negated expressions.)
4. $H \Rightarrow S$
5. $S \Leftrightarrow (\neg H)$

Now, as things stand, we can represent simple propositions by letters, and combine them together using the connectives, with brackets to ensure that there is no ambiguity in the result. However, this can lead to expression with many levels of brackets, which reduces readability. One way of obviating this problem is to have precedence rules, rather like those for arithmetic. (If we see the expression $2 + 3 \times 4$ we know that we do the multiplication first, then the addition.)

The rules of precedence are as follows:

1. Negation binds most tightly, so that $\neg P \oplus Q$ means $(\neg P) \oplus Q$, whichever connective $\oplus$ is.
2. Disjunction and conjunction bind next most tightly, and are of equal strength. So $P \vee Q \Rightarrow \neg R$ means $(P \vee Q) \Rightarrow (\neg R)$, but brackets are needed to say whether $P \vee Q \wedge R$ means $(P \vee Q) \wedge R$ or $P \vee (Q \wedge R)$.
3. Implication is next.
4. Equivalence is weakest.

It is possible to introduce further rules that reduce further the number of brackets required to make an expression unambiguous, but this can make it harder to read and interpret the expression. The conventions used here give a good compromise between readability and conciseness.

Worked examples

1.9 Remove as many brackets as you can from

1. $(\neg P) \Rightarrow ((P \wedge Q) \vee R)$
2. $\neg(P \Rightarrow ((Q \vee R) \wedge (\neg P)))$
3. $P \Rightarrow (Q \Rightarrow R)$
4. $(P \wedge Q) \Leftrightarrow (P \vee Q)$

Solution

1. $\neg P \Rightarrow (P \wedge Q) \vee R$
2. $\neg(P \Rightarrow (Q \vee R) \wedge \neg P)$
3. $P \Rightarrow (Q \Rightarrow R)$
4. $P \wedge Q \Leftrightarrow P \vee Q$

1.10 Introduce brackets to make the following expressions unambiguous to someone unfamiliar with the precedence rules.

1. $\neg P \vee Q \Rightarrow P \wedge \neg Q$
2. $P \wedge \neg S \Rightarrow Q \vee R$

Solution

1. $((\neg P) \vee Q) \Rightarrow (P \wedge (\neg Q))$
2. $(P \wedge (\neg S)) \Rightarrow (Q \vee R)$

Finally, there is some useful terminology relevant to the implication connective. If $P \Rightarrow Q$, we say that P *implies* Q. However there are several alternative ways of expressing the same idea.

The most obvious English sentence that may be rendered $P \Rightarrow Q$ is '*if P, then Q*', but there are others. If you know that '*P* is a *sufficient condition* for *Q*', then you know that whenever *P* is true, *Q* must also be true, so we can also represent this by $P \Rightarrow Q$.

It is also possible to express the same thing in what is perhaps a less familiar way; if you know that *Q* is a *necessary* condition for *P*, then for *P* to be true it must be the case that *Q* is true, so again we have $P \Rightarrow Q$; similarly, if you know that *P* is true only when *Q* is true, the truth of *P* enforces that of *Q*, so we have $P \Rightarrow Q$.

It is important to understand and remember the distinction between **necessary** and **sufficient** conditions.

Finally, if *P* and *Q* are logically equivalent, we say that *P* is a *necessary and sufficient* condition for *Q* (or that *Q* is a necessary and sufficient condition for *P*), or that *P* is true *if and only if Q* is true.

Self-Assessment Questions 1.2

Each of the following sentences may be represented by either $P \Rightarrow Q$ or $Q \Rightarrow P$. Decide which is appropriate in each case, and give your reasons.

1. If *P* is true, then *Q* is true.
2. If *P* is false, then *Q* is false.
3. *Q* is only true if *P* is true.
4. *Q* is a sufficient condition for *P*.
5. *Q* is a necessary condition for *P*.

Exercise 1.2

1. Write the conjunction of the following two propositions using the new symbol.

   (a) My program compiled without any problems. (b) My program had run-time errors.

2. Write down the disjunction of the statements

   (a) The specification is oversimplified.
   (b) The programming team is unhappy with the specification.

   What can you deduce from the fact that the programming team is happy with the specification?

3. If the proposition 'the specification is suitable $\Rightarrow$ the programming team is happy' is true, could it be the case that the programming team is unhappy but the specification is nevertheless suitable?

4. If the proposition 'The specification is adequate $\Leftrightarrow$ the programming team is unhappy' is true, can it be the case that the specification is adequate and the programming team is happy?

5. Give the negation of each of the following, both in normal English and using the $\neg$ symbol.

(a) The programming team is happy.

(b) The programming team is always happy.

(c) $S \vee (\neg H)$ (d) $(S \Rightarrow H) \vee (S \wedge (\neg H))$

(e) $S \Leftrightarrow H$

6. Write out the English version of the following propositions, using the same identifications between letters and simple propositions as in Worked example 1.8.

(a) $(\neg S) \wedge H$ (b) $(\neg H) \Rightarrow S$

7. Remove all unnecessary brackets from

(a) $((\neg P) \vee Q) \Rightarrow (P \wedge (\neg Q))$

(b) $(P \Rightarrow (\neg Q)) \Leftrightarrow (Q \vee (P \wedge (\neg Q)))$

1.3 Truth tables: construction

We now have a concise way of expressing complicated propositions, and combinations of them. Using this, we can build up a collection of tools for analysing statements, arguments, and so on. One approach is to consider propositions in terms of their truth or falsity, which is determined by the truth or falsity of the simple propositions comprising them.

For example, if we consider the proposition $P \wedge Q$, we see that it is true if both P and Q are true, and false otherwise. But although we can readily determine the truth value of a proposition like $P \wedge Q$, things can easily get much more complicated. For this reason, it is necessary to develop a systematic way of considering the truth value of a complicated expression. The technique we will consider here is that of *truth tables*, which give us a way of listing the truth value of a complicated proposition in terms of the truth values of the simple propositions comprising it.

Recall that a true proposition has truth value T, and a false one has truth value F. First, we develop truth tables for the standard connectives; then we see how to combine these into truth tables for more complicated propositions.

The basic truth tables

Let P and Q be any simple propositions. Then we have the truth tables shown in Figure 1.1. The first three of these truth tables are reasonably obvious; the table for implication (table (d)), however, occasionally causes some confusion. The first two lines are plausible. They say that if you start off with a true statement, you can only deduce a true statement. But what do the last two tell us? We can interpret the last two lines as saying that if you start off with false assumptions, you can end up with any result, whether it is true or false. One way of seeing this is to keep firmly in mind that implication is supposed to be a way of getting to a correct conclusion from correct assumptions. From this point of view, the only time an implication is false is if we start off with something true, and end up with something false. The truth table shown does this job adequately, and is completely standard, so we will use it.

Truth tables for general propositions

Now, using these truth tables as building blocks, we can construct truth tables for any proposition. The main point to bear in mind is that we must be sure not to miss any lines out, and that we must be sure to stick to the rules of precedence.

(a)

P	$\neg P$
T	F
F	T

(b)

P	Q	$P \vee Q$
T	T	T
T	F	T
F	T	T
F	F	F

(c)

P	Q	$P \wedge Q$
T	T	T
T	F	F
F	T	F
F	F	F

(d)

P	Q	$P \Rightarrow Q$
T	T	T
T	F	F
F	T	T
F	F	T

(e)

P	Q	$P \Leftrightarrow Q$
T	T	T
T	F	F
F	T	F
F	F	T

Figure 1.1

Worked example

1.11 Construct the truth tables for

1. $P \vee \neg Q$
2. $\neg(\neg P \wedge \neg Q)$
3. $P \wedge Q \Rightarrow R$

Solution We build up the result column by column in the truth tables. Each step of the first example is shown, but thereafter, only the final result is given, and the numbers in the bottom row of each truth table give the order in which the columns are calculated.

1. First, we fill in the columns which simply repeat the simple propositions P and Q, as shown in Figure 1.2(a). Next, we evaluate those columns that depend only on the ones calculated so far; in this case we include column 3, as shown in Figure 1.2(b). Finally, we fill in the column that depends on columns 1 and 3, namely column 4; the final truth table is in Figure 1.2(c).

P	Q	P	$\vee$	$\neg$	Q
T	T	T			T
T	F	T			F
F	T	F			T
F	F	F			F
		1			2

(a)

P	Q	P	$\vee$	$\neg$	Q
T	T	T		F	T
T	F	T		T	F
F	T	F		F	T
F	F	F		T	F
		1		3	2

(b)

P	Q	P	$\vee$	$\neg$	Q
T	T	T	T	F	T
T	F	T	T	T	F
F	T	F	F	F	T
F	F	F	T	T	F
		1	4	3	2

(c)

Figure 1.2

2. Henceforth, only the final version of each truth table will be shown. You should go through the construction of each example, column by column, until you are confident in your ability to carry out the procedure. Column 6 of Figure 1.3 shows the truth values for $\neg(\neg P \wedge \neg Q)$.

Figure 1.3

P	Q	$\neg$	$(\neg$	P	$\wedge$	$\neg$	$Q)$
T	T	T	F	T	F	F	T
T	F	T	F	T	F	T	F
F	T	T	T	F	F	F	T
F	F	F	T	F	T	T	F
		6	3	1	5	4	2

Note that in this case the final result is identical to that for $P \vee Q$, so we see that $P \vee Q \Leftrightarrow \neg(\neg P \wedge \neg Q)$.

3. Column 5 of Figure 1.4 gives the truth values in this case.

P	Q	R	P	$\wedge$	Q	$\Rightarrow$	R
T	T	T	T	T	T	T	T
T	T	F	T	T	T	F	F
T	F	T	T	F	F	T	T
T	F	F	T	F	F	T	F
F	T	T	F	F	T	T	T
F	T	F	F	F	T	T	F
F	F	T	F	F	F	T	T
F	F	F	F	F	F	T	F
			1	4	2	5	3

Figure 1.4

Note carefully the pattern of Ts and Fs in the first columns of the truth tables. The pattern is identical to the pattern of 0s and 1s in counting using binary arithmetic. It guarantees that all necessary rows will appear in the truth table, and is recommended most highly as the method of choice for constructing these tables.

Self-Assessment Question 1.3

Does the construction of truth tables do anything for us that we could not do just by arguing about the truth values of the simple propositions comprising a compound proposition?

Exercise 1.3

1. By constructing truth tables, show that $P \vee (Q \vee R) \Leftrightarrow (P \vee Q) \vee R$ and $P \wedge (Q \wedge R) \Leftrightarrow (P \wedge Q) \wedge R$.
2. Show that brackets are required for the interpretation of $P \vee Q \wedge R$, *i.e.* show that $P \vee (Q \wedge R)$ is not logically equivalent to $(P \vee Q) \wedge R$.

1.4 Truth tables: applications

Given competence in constructing truth tables we can begin to carry out useful tasks using them. For example, we can study an argument and see if it is **logically valid**, *i.e.* see whether the supposed consequence really does follow from the antecedent. We can examine a collection of statements and see whether they are **consistent**, *i.e.* see whether it is possible for them all to be true at the same time. It is important, for example, for a specification to be consistent.

Before carrying on with these goals, however, it is convenient to have a little more terminology. A proposition is said to be a **tautology** if it always has the truth value T, a **contingency** if it sometimes has truth value T and sometimes F, and an **inconsistency** if it always has the truth value F. A

consistent proposition is one which is a tautology or a contingency. Clearly, we can establish which of these classes a proposition falls into just by constructing its truth table.

Worked example

1.12 Classify each of the following as tautologous, contingent, or inconsistent.

1. $P \wedge \neg P$ 2. $P \vee \neg P$ 3. $P \vee Q$

Solution The truth tables for each of these are shown in Figures 1.5, 1.6 and 1.7 respectively.

P	$\neg P$	$P \wedge \neg P$
T	F	F
F	T	F

Figure 1.5

P	$\neg P$	$P \vee \neg P$
T	F	T
F	T	T

Figure 1.6

P	Q	$P \vee Q$
T	T	T
T	F	T
F	T	T
F	F	F

Figure 1.7

1. Since the final column of the truth table in Figure 1.5 consists entirely of Fs, the statement is inconsistent.
2. The final column of Figure 1.6 consists entirely of Ts, so the statement is a tautology.
3. The final column of Figure 1.7 has both Ts and Fs, hence the statement is contingent.

Consistency

A collection of propositions P, Q, ..., R is consistent if $P \wedge Q \wedge \ldots \wedge R$ is consistent. To establish the consistency of a collection of propositions, we simply construct the truth table of their conjunction. If the final column of this truth table has any Ts in it, the statements are consistent.

Worked example

1.13 Is the following collection of statements consistent?

1. If Agatha is on the programming team, Bert is on it.
2. Agatha and Bert are on the programming team.
3. If Cynthia is on the programming team, Bert isn't.

Solution First, we identify the simple propositions out of which each of these is built. It is easy to see that 'Agatha is on the programming team', 'Bert is on the programming team' and 'Cynthia is on the programming team' will do. Next, we need letters to represent each of these propositions. Fortuitously, the people involved have names beginning with the first three letters of the alphabet, so we use A, B and C respectively to represent our three simple propositions. Then we want to check the following compound proposition for consistency:

$$(A \Rightarrow B) \wedge (A \wedge B) \wedge (C \Rightarrow \neg B)$$

(Strictly speaking, we do not require all the brackets used here, but they help readability.) The truth table then becomes that shown in Figure 1.8. Column 12 contains a T, and so the collection of statements is consistent.

Figure 1.8

A	B	C	$(A$	$\Rightarrow$	$B)$	$\wedge$	$(A$	$\wedge$	$B)$	$\wedge$	$(C$	$\Rightarrow$	$\neg$	$B)$
T	T	T	T	T	T	T	T	T	T	F	T	F	F	T
T	T	F	T	T	T	T	T	T	T	T	F	T	F	T
T	F	T	T	F	F	F	T	F	F	F	T	T	T	F
T	F	F	T	F	F	F	T	F	F	F	F	T	T	F
F	T	T	F	T	T	F	F	F	T	F	T	F	F	T
F	T	F	F	T	T	F	F	F	T	F	F	T	F	T
F	F	T	F	T	F	F	F	F	F	F	T	T	T	F
F	F	F	F	T	F	F	F	F	F	F	F	T	T	F
			1	8	2	10	3	9	4	12	5	11	7	6

Valid arguments

We can also use truth table arguments to establish the validity of an argument. To see whether the proposition P is a consequence of the premisses A, B, $\ldots C$, we simply form the proposition $A \wedge B \wedge \ldots \wedge C \Rightarrow P$ and check by constructing its truth table whether it is a tautology. If it is, then whenever the premisses are all true, P is also true, and so the argument is valid. If this is the case, we also say that $A \wedge B \wedge \ldots \wedge C$ **logically implies** P.

Worked example

1.14 Is the following argument valid?
If my program is correct, it will pass the validation test. My program passes the validation test. Therefore, my program is correct.

Solution Again, we begin by setting up the argument in symbolic form. Using C to represent the simple proposition 'My program is correct', and V to represent 'My program passes the validation test', we can rewrite the above argument as $(C \Rightarrow V) \wedge V \Rightarrow C$, and this is the proposition that must be a tautology for the argument to be valid. We next construct the truth table, shown in Figure 1.9. So we see from column 7 that the proposition is not a tautology, and so the argument is not in fact valid; it is possible for an incorrect program to pass the validation test.

C	V	$(C$	$\Rightarrow$	$V)$	$\wedge$	V	$\Rightarrow$	C
T	T	T	T	T	T	T	T	T
T	F	T	F	F	F	F	T	T
F	T	F	T	T	T	T	F	F
F	F	F	T	F	F	F	T	F
		1	5	2	6	3	7	4

Figure 1.9

If the proposition $P_1 \wedge \ldots \wedge P_n \Rightarrow Q$ is a tautology, we also write

$$P_1 \wedge \ldots \wedge P_n \models Q$$

which is pronounced 'P_1 to P_n **semantically entail** Q'. (This means that by checking all the possibilities we can show that whenever P_1 to P_n are true, Q is also true.)

Self-Assessment Question 1.4

Why is the statement '$P_1 \wedge \ldots \wedge P_n \Rightarrow Q$ is a tautology' equivalent to the statement '$P_1 \wedge \ldots \wedge P_n \wedge \neg Q$ is a contradiction'?

Exercise 1.4

1. (a) Show that $P \Rightarrow Q$ is logically equivalent to $\neg P \vee Q$.
   (b) Is $P \Rightarrow Q$ logically equivalent to $Q \Rightarrow P$?
2. (a) Rewrite the proposition $P \wedge Q \Rightarrow P$ without using the implication sign. Show that the proposition is a tautology.
   (b) Show that $P \Rightarrow Q \Leftrightarrow \neg(P \wedge \neg Q)$ is a tautology.

1.5 Algebraic laws

In the exercises above we saw that not all of the connectives are necessary, since we can rewrite some of them in terms of others. The equivalences were shown using truth tables. We can now use truth tables to demonstrate the equivalence of some collections of expressions. However, the point of these equivalences is not to enable us to do away with some of the connectives – it is to establish a useful working set of algebraic laws which can be used to re-express propositions in useful ways. For propositions comprising more than a few simple propositions this can be a much more efficient way of proceeding than trying to construct truth tables with dozens or even hundreds of lines.

The following equivalences are all established simply by constructing truth tables, but the important thing is to develop familiarity with them and a facility for using them to manipulate propositions, rather than to spend inordinate periods of time drawing up truth tables to check them all.

Think of these laws as being analogous to the laws of algebra used when analysing an expression where the letters represent unknown numbers; we know that any expression of the form $a(b+c)$ may be rewritten as $ab+ac$, where a, b and c are numbers (which may in turn be expressed as complicated expressions). In the same way, the following laws allow us to re-express propositions. P, Q and R represent arbitrary propositions.

KEY POINT

1. Commutativity

$$P \vee Q \Leftrightarrow Q \vee P \qquad P \wedge Q \Leftrightarrow Q \wedge P$$

2. Associativity

$$P \vee (Q \vee R) \Leftrightarrow (P \vee Q) \vee R \qquad P \wedge (Q \wedge R) \Leftrightarrow (P \wedge Q) \wedge R$$

3. Distributivity

$$P \vee (Q \wedge R) \Leftrightarrow (P \vee Q) \wedge (P \vee R)$$
$$P \wedge (Q \vee R) \Leftrightarrow (P \wedge Q) \vee (P \wedge R)$$

4. Idempotency

$$P \vee P \Leftrightarrow P \qquad P \wedge P \Leftrightarrow P$$

5. de Morgan's Laws

$$\neg(P \vee Q) \Leftrightarrow \neg P \wedge \neg Q \qquad \neg(P \wedge Q) \Leftrightarrow \neg P \vee \neg Q$$

6. Properties of negation

$$\neg\neg P \Leftrightarrow P \quad P \vee \neg P \Leftrightarrow T \quad P \wedge \neg P \Leftrightarrow F \quad \neg T \Leftrightarrow F \quad \neg F \Leftrightarrow T$$

7. Properties of T and F

$$P \vee T \Leftrightarrow T \quad P \wedge T \Leftrightarrow P \quad P \vee F \Leftrightarrow P \quad P \wedge F \Leftrightarrow F$$

This provides us with a new way of approaching problems involving propositions. Rather than constructing the truth table for $\mathcal{P} \Rightarrow Q$ (where $\mathcal{P}$ is a compound proposition), and checking whether it is a tautology, we can use the algebraic laws and see whether it can be manipulated directly to T. If it is possible to replace the original expression by an equivalent expression using the laws given above, then that expression by an equivalent one, and so on, ending up with just T, then the original expression must be a tautology. Likewise, to see whether a proposition is a contradiction, we can try to manipulate it to F. At the very least, we may be able to simplify the expression so that its nature is obvious.

Note that none of the laws given involve the implication connective; however, since $P \Rightarrow Q$ is logically equivalent to $\neg P \vee Q$, we can use this fact to re-express any proposition in a way that involves only negation, conjunction and disjunction, and we do not have to introduce new laws.

Worked examples

1.15 Is the following argument valid?

If my program is correct, then it will pass the validation tests. My program is correct. Therefore, my program will pass the validation tests.

Solution As before, we use C to represent the statement 'My program is correct', and V to represent 'My program will pass the validation tests'. Then we can write the argument as $(C \wedge (C \Rightarrow V)) \Rightarrow V$. We can rewrite this as $(C \wedge (\neg C \vee V)) \Rightarrow V$, and then as $\neg(C \wedge (\neg C \vee V)) \vee V$, and then use the algebraic laws to test the validity of the argument.

$$\neg(C \wedge (\neg C \vee V)) \vee V \Leftrightarrow \neg C \vee \neg(\neg C \vee V) \vee V \quad \text{(de Morgan)}$$
$$\Leftrightarrow \neg C \vee (C \wedge \neg V) \vee V \quad \text{(de Morgan)}$$
$$\Leftrightarrow ((\neg C \vee C) \wedge (\neg C \vee \neg V)) \vee V \quad \text{(distributive law)}$$
$$\Leftrightarrow (T \wedge (\neg C \vee \neg V)) \vee V \quad \text{(property of } T\text{)}$$
$$\Leftrightarrow \neg C \vee \neg V \vee V \quad \text{(property of } T\text{)}$$
$$\Leftrightarrow \neg C \vee T \quad \text{(property of } T\text{)}$$
$$\Leftrightarrow T \quad \text{(property of } T\text{)}$$

It follows that the proposition $(C \wedge (C \Rightarrow V)) \Rightarrow V$ is a tautology, and therefore that the argument is valid.

Note: this solution has been written out in more detail than one would normally bother with, to show just how the laws are being used. In practice, fewer lines would almost certainly be used.

1.16 Show that $(P \Rightarrow Q) \wedge (Q \Rightarrow R) \Rightarrow (P \Rightarrow R)$ is a tautology.

Solution The approach is, as always, first to rewrite the expression without the use of the $\Rightarrow$ sign, then to manipulate the resulting expression using the algebraic laws. This yields

$$\begin{aligned}
&\neg((\neg P \vee Q) \wedge (\neg Q \vee R)) \vee (\neg P \vee R) \\
\Leftrightarrow &(P \wedge \neg Q) \vee (Q \wedge \neg R) \vee \neg P \vee R \\
\Leftrightarrow &(P \wedge \neg Q) \vee \neg P \vee R \vee (Q \wedge \neg R) \\
\Leftrightarrow &(P \wedge \neg Q) \vee \neg P \vee R \vee Q \\
\Leftrightarrow &\neg Q \vee \neg P \vee R \vee Q \\
\Leftrightarrow &T
\end{aligned}$$

where less detail has been given. Since the original expression is logically equivalent to T, it is a tautology.

Self-Assessment Question 1.5

Explain how to use the algebraic laws to show that an argument is valid.

Exercise 1.5

1. Is the following argument valid?

   If the programming team is happy, the specification is clear.
   The specification is not clear. Therefore the programming team is not happy.

2. Show that the following set of statements is inconsistent.

   If John is on the programming team, Alice is not on the programming team. Fred and Alice are both on the programming team. If Fred is on the programming team, John is on the programming team.

1.6 Deduction and proof

Between the truth tables approach and the algebraic one, we have two useful ways of attacking problems involving propositions. There is one more important technique which we will consider, which is of particular relevance in constructing valid arguments. The truth tables and the algebraic laws we have are most useful for checking logical equivalence of two statements, or checking that a statement is either a tautology or an inconsistency. There is another method that has a rather different strength. It is a method of starting off with some assumptions, and constructing from those assumptions a chain of propositions such that the final proposition in the chain logically follows from the assumptions. Such a chain of propositions is what is called a **proof** in mathematical logic, and the final proposition is called a **theorem**.

The fundamental idea of this is to have a collection of valid argument types, and to use these to obtain conclusions that logically follow from the assumptions. We will consider one particular approach to this; there are many others. These valid argument types are called **rules of inference**, and are generally represented as tableaux. Each tableau consists of a collection of propositions, called the **hypotheses** or **premisses**, followed by a horizontal line and then by the **conclusion**. For example, a well-known rule of inference is the *modus ponens*, which may be represented by

$$\frac{\begin{array}{c}P\\ P \Rightarrow Q\end{array}}{Q}$$

That this argument form is valid follows from the fact that $P \wedge (P \Rightarrow Q) \Rightarrow Q$ is a tautology. Each of the rules of inference corresponds to some tautology involving implication; the point of this method is to use these tautologies to construct proofs rather than having to construct truth tables for propositions comprising many simple propositions.

The rules of inference

We will use the following list of nine rules of inference; there are systems which use more, and systems which use fewer. The system presented here is a fairly common one in which it is relatively easy to construct proofs.

KEY POINT

1. Addition $\dfrac{P}{P \vee Q}$
2. Simplification $\dfrac{P \wedge Q}{P}$
3. *Modus ponens* $\dfrac{\begin{array}{c}P\\ P \Rightarrow Q\end{array}}{Q}$
4. *Modus tollens* $\dfrac{\begin{array}{c}\neg Q\\ P \Rightarrow Q\end{array}}{\neg P}$
5. Disjunctive syllogism $\dfrac{\begin{array}{c}P \vee Q\\ \neg P\end{array}}{Q}$
6. Hypothetical syllogism $\dfrac{\begin{array}{c}P \Rightarrow Q\\ Q \Rightarrow R\end{array}}{P \Rightarrow R}$
7. Conjunction $\dfrac{\begin{array}{c}P\\ Q\end{array}}{P \wedge Q}$

8. Constructive dilemma

$$\frac{\begin{array}{c}(P \Rightarrow Q) \wedge (R \Rightarrow S)\\ P \vee R\end{array}}{Q \vee S}$$

9. Destructive dilemma

$$\frac{\begin{array}{c}(P \Rightarrow Q) \wedge (R \Rightarrow S)\\ \neg Q \vee \neg S\end{array}}{\neg P \vee \neg R}$$

Although the names of most of these are English and self-explanatory, there are two in Latin. The first is called *modus ponens*, which means 'method of putting'; one 'puts' the premiss of the implication, and draws the conclusion. The second is called *modus tollens*, which means 'method of taking'; one 'takes' the consequence of the implication, and deduces that the premiss must be false. It is useful to use the names of these rules, rather than the numbers, because the names are standard, while different books may give them in different orders.

We note that the rules of inference all have one thing in common, namely that the conjunction of the hypotheses logically implies the conclusion. This ensures that if the hypotheses are true, so also will be the conclusion, which is just what we want.

Worked example

1.17 Show that the conjunction of the hypotheses of the disjunctive syllogism logically implies the conclusion.

Solution The conjunction of the hypotheses is $(P \vee Q) \wedge \neg P$, which is logically equivalent to $\neg P \wedge Q$. We thus obtain

$$\begin{aligned}(P \vee Q) \wedge \neg P \Rightarrow Q &\Leftrightarrow \neg(\neg P \wedge Q) \vee Q\\ &\Leftrightarrow P \vee \neg Q \vee Q\\ &\Leftrightarrow T\end{aligned}$$

Constructing proofs

Using the rules of inference to construct proofs is not a process for which it is easy to provide an algorithm that can be used efficiently by humans. Although there are algorithms which enable computers to carry out formal reasoning, they tend to rely heavily on the ability to make large numbers of attempts quickly. People must usually use imagination and insight, which are strengthened by practice; the best way to improve is by doing examples.

One lays out a proof as a table. Each line consists of a hypothesis, or of a proposition that follows from previous lines by means of a rule of inference, or by using one of the logical equivalences from the algebraic laws. It is normal to list the justification for each step in a column parallel to the column of propositions. The consequence is the last line of the proof.

Worked example

1.18 Construct a proof to show that the following argument is valid.

> If the programming team is happy, or the specification is clear, then the program will be written under budget. If the program is written under budget, the boss will rejoice. The boss will not rejoice. Therefore, the specification is not clear.

Solution First, we decide upon letters to represent the propositions. We will use the following:

H denotes 'The programming team is happy'
C denotes 'The specification is clear'
B denotes 'The job will be done under budget'
R denotes 'The boss will rejoice'.

Then the hypotheses are $H \vee C \Rightarrow B$, $B \Rightarrow R$, and $\neg R$. We wish to prove from these hypotheses the conclusion $\neg C$.

Proof:

	Assertion	Reason
1.	$H \vee C \Rightarrow B$	Hypothesis 1
2.	$B \Rightarrow R$	Hypothesis 2
3.	$H \vee C \Rightarrow R$	1, 2, hypothetical syllogism
4.	$\neg R$	Hypothesis 3
5.	$\neg(H \vee C)$	3, 4, *modus tollens*
6.	$\neg H \wedge \neg C$	5, de Morgan's law
7.	$\neg C \wedge \neg H$	6, commutativity of $\wedge$
8.	$\neg C$	7, simplification

So we see that the conclusion does indeed follow from the hypotheses.

If it is possible to construct a proof of Q from the hypotheses $P_1 \ldots P_n$, then we write

$$P_1, \ldots, P_n \vdash Q$$

which is pronounced 'P_1 to P_n **syntactically entail** Q'. This phrase means that the truth of Q can be shown to follow from the algebraic form of the statements P_1 to P_n, without reference to their actual truth values. It is obvious that if $P_1, \ldots, P_n \vdash Q$ then $P_1, \ldots, P_n \models Q$ since each rule of inference can only take us from true propositions to true ones. It is less obvious that if $P_1, \ldots, P_n \models Q$ then $P_1, \ldots, P_n \vdash Q$, because we may not have enough rules of inference. It can, however, be shown that the rules of inference listed previously are sufficient. Thus semantic and syntactic entailment are equivalent in the propositional calculus.

Self-Assessment Question 1.6

How might you use a method of constructing logically valid arguments to prove that two propositions are logically equivalent? (*Hint*: use a truth table to show that $P \Leftrightarrow Q$ is logically equivalent to $(P \Rightarrow Q) \wedge (Q \Rightarrow P)$.)

Exercise 1.6

1. Show that the conjunction of the hypotheses of *modus tollens* logically implies the conclusion.
2. Construct a proof for the following argument.

   If today is the deadline, then we must finish the specification or start the program. If the team leader is unhappy, we do not start the program. Today is the deadline and the team leader is unhappy. Therefore we must finish the specification.

Test and Assignment Exercises 1

1. Use truth tables to establish each of the algebraic laws of propositional calculus.
2. Argue that $P \Rightarrow Q$ should be equivalent to $\neg(P \wedge \neg Q)$, and so justify the truth table for implication.
3. Find an expression for $P \wedge Q$ involving only P, Q, $\neg$ and $\vee$.
4. Use (i) truth tables and (ii) algebraic manipulation to establish the equivalence of each of the following pairs:

   (a) $(P \vee Q) \wedge \neg P$ and $Q \wedge \neg P$ (b) $P \Rightarrow (Q \Rightarrow P)$ and T (c) $(P \Rightarrow (Q \Rightarrow R)) \Rightarrow ((P \Rightarrow Q) \Rightarrow (P \Rightarrow R))$ and T (d) $P \wedge (P \vee Q)$ and P (e) $P \vee (P \wedge Q)$ and P
5. Use (i) truth tables and (ii) algebraic manipulation to show that the consequence of each of the rules of inference is logically implied by the conjunction of its hypotheses.
6. Show that $(P \Leftrightarrow F) \Leftrightarrow \neg P$.
7. In a certain medical practice, the data on patients are held on computer. In an attempt to keep data secure and at the same time stay within the requirements of the practice's policy of making a patient's data available to that patient, the following list of requirements is drawn up for files in the practice's database.

   (a) All the files in the database are encrypted.

   (b) A file is not encrypted unless it contains confidential information.

   (c) If a file is accessible to patients, then it must be possible to copy it.

   (d) Files containing confidential information are covered by the practice's policy of making a patient's data available to the patient.

(e) If a file is encrypted, then it must not be possible to copy it.

(f) If a file is covered by the practice's policy of making a patient's data available to the patient, then it must be accessible to patients.

Given this specification,

(i) Use algebraic manipulation to show that this specification is inconsistent if there are any files at all in the database. (*Hint*: pick some file, and choose letters to represent simple propositions about it, then combine them into a compound proposition using the above specification.)

(ii) Use the rules of inference to show that the specification is inconsistent if there are any files at all in the database. (*Note*: this is equivalent to showing that from the specification you can deduce F, or $P \wedge \neg P$ for some proposition P.)

8. Are the following statements consistent?

   The system manager was incompetent or, if the system crashed then that day's log entries are complete. If the system manager was competent, the system did not crash. If the log entries are complete, then the system crashed.

9. Show that the following argument is valid.

   If we computerize our stock control, our rival companies will feel threatened and our own employees will become complacent. If our rival companies feel threatened, they may start a price war. If our own employees become complacent, they will put less effort into sales. If we do not computerize our stock control, then we are very likely to go bankrupt if there is a price war. Therefore, our rival companies may start a price war and our own employees will put less effort into sales, or we are very likely to go bankrupt if there is a price war.

1.7 Further reading

Lewis Carroll, *Symbolic Logic and The Game of Logic* (Dover, 1958)
This book uses board games to introduce techniques of formal reasoning.

W P Hodges, *Logic* (Penguin, 1977)
A popular book on logic which discusses some of the more technical issues in an accessible way.

A Kaldewaij, *Programming: The Derivation of Programs* (Prentice Hall International, 1990)
This is a programming text which relies heavily on techniques of formal logic to produce reliable programs.

A Margaris, *First Order Mathematical Logic* (Dover, 1990)
A standard text on mathematical logic.

E Mendelson, *Introduction to Mathematical Logic* (Wadsworth & Brooks/Cole, 1987)
Another standard text on mathematical logic, with some applications to computing science.

R Smullyan, *Satan, Cantor, and Infinity* (Oxford University Press, 1993) and others
These are books which use puzzles as a medium for teaching the principles of logic, starting off with simple logic puzzles and leading on to sophisticated topics.

2 Sets

Objectives

When you have mastered this chapter you should be able to

- calculate unions, intersections, differences and Cartesian products of simple sets
- use tabular and diagrammatic methods to analyse set algebraic expressions
- use algebraic methods to manipulate set algebraic expressions
- use results about the number of elements of the union and intersection of sets
- understand functions as rules associating elements of one set to elements of another, and classify them as injective, surjective or bijective when they satisfy the appropriate conditions
- know what a relation is in terms of Cartesian products, and what the connection is between relations and functions

Now that we have a formal way of arguing and checking proofs, we can develop some objects about which we can argue. The objects we will consider are sets, which are, naively, collections. Set theory is important in the foundations of mathematics. However, we will be concentrating on more practical aspects, rather than theoretical ones. For example, programs may be thought of as functions from one set of data to another; the use, understanding and construction of formal specifications requires the use of sets; the concepts that underlie relational databases are set-algebraic; and probability, which is applied, for example, to risk analysis, uses the language of set theory in a fundamental way. Before we are in a position to see how sets may be used in these circumstances, though, there is a certain amount of spade-work to be done.

2.1 What is a set?

Roughly speaking, a **set** is any collection; the objects of the collection are called its **elements**. If a set consists only of a small number of items, we can represent the set simply by listing its elements between braces. For example,

$$A = \{a,\ 2,\ \text{red}\}$$

describes a set, A, whose elements are the letter a, the number 2, and the colour red. We use the notation

$$a \in A$$

to say that a is an element of A. The order in which the items of a set are listed is unimportant; we could equally well say

$$A = \{\text{red},\ 2,\ a\}$$

or put the elements in any order we liked. It is also unimportant how often an element is listed – all that matters is which elements are listed, not the order or the frequency. So we could also write

$$A = \{\text{red},\ 2,\ a,\ 2,\ 2,\ \text{red}\}$$

We can consider another set, $B = \{a,\ 2,\ \text{red},\ \text{green}\}$. All the elements of A are also elements of B, so we say that A is a **subset** of B and write $A \subseteq B$. Two sets X and Y are equal if they have exactly the same elements, *i.e.* $X \subseteq Y$ and $Y \subseteq X$. In the case of A and B here, A is a subset of B, but A is not equal to B. In this case, we say that A is a **strict subset** of B, which we denote by $A \subset B$. The set with no elements in it at all is called the **empty set**, and is denoted by $\{\ \}$, or $\emptyset$. We will use the latter notation here. By convention, if A is any set, we say that $\emptyset \subseteq A$ (since there are never elements in $\emptyset$ that are not in A).

Worked example

2.1 Which of the following sets are equal to the others, and which are subsets of the others?

$$A = \{1,2,3\} \quad B = \{1,3\} \quad C = \{1,2,4\} \quad D = \{1,2,3,4\}$$
$$E = \{2,1,3\}$$

Solution $A = E,\ A \subset D,\ A \subseteq E,\ B \subset A,\ B \subset D,\ B \subset E,\ C \subset D,\ E \subseteq A,\ E \subset D.$

At first sight, this seems like a very straightforward idea; it is clear that some more efficient way of describing big sets will be needed, but it doesn't seem as though anything can go really wrong. This first impression is, however, misleading. So, before we go on to do anything with sets, we will discuss a couple of relatively subtle issues.

The first point is that an element of a set may be a set in its own right. For example, the set

$$B = \{a, \{1, 2\}, \{\{b\}\}\}$$

has elements a, $\{1, 2\}$ and $\{\{b\}\}$.

Worked example

2.2 What set has elements a, $\{\{1\}, 2\}$ and 1?

Solution $\{a, \{\{1\}, 2\}, 1\}$.

The second point is a genuine problem for this naive approach to set theory. If we allow arbitrary collections as sets, then we do not have a consistent theory. This was famously observed by Bertrand Russell and is demonstrated in the paradox that carries his name.

Russell's Paradox Let P be the collection of all sets A satisfying $A \in A$. If we allow arbitrary collections as sets, then there are such sets, since, for example, the collection of all objects that are not my left foot is an element of itself. Next, let Q be all the other sets, so that Q consists of all those sets A that are *not* elements of themselves, which we write $A \notin A$. (Note that $x \notin A$ is just another way of writing $\neg(x \in A)$.) We now ask, what kind of set is Q? Is it an element of P or of Q? Since between the two we have included all collections, it must be in one or the other. But if $Q \in Q$, it satisfies the condition for being in P, which means that $Q \notin Q$. On the other hand, if $Q \notin Q$, this means – by the very definition of Q! – that $Q \in Q$. There is something terribly wrong here. It would obviously be better to have a consistent notion of what a set is.

If you find the previous paragraph confusing, there is an analogy which may help. Think of a library, which contains two catalogues, which we will call catalogue P and catalogue Q. Catalogue P is a list of all those books which have some mention of themselves, and catalogue Q lists all those which do not. Now, it is easy to see that P is to be listed in catalogue P. But what about Q? If we put it in Q, it mentions itself, and so it ought to be in P. But if we list it in P, it is not mentioned in Q, and so it ought to be in catalogue Q. Either way, Q is listed in the wrong catalogue. If you think of 'is listed in' as corresponding to 'is an element of', you should be able to get some idea of the original problem.

In fact, the root of the problem is that we are assuming that there is a set of all sets. We can avoid contradictions such as the above by always working in a particular well-defined domain of discourse, or **universal set**; for example, if we are considering computations, we might be concerned with all possible integers, or even all sets of integers, and so on. In fact, in those problems involving sets which we will be concerned with, we will usually regard it as obvious what the universal set is, and not even bother to state it explicitly;

however, you should be aware that there are problems with the notion of set, which will reappear if you choose to study such subjects as the theory of computability in detail.

Self-Assessment Question 2.1

Explain why it is necessary in principle to have a universal set for a given problem.

Exercise 2.1

What are the elements of $\{1, \{1\}, \{1, \{1\}\}\}$?

2.2 New sets from old

From now on, then, we will assume that all the sets we are interested in are subsets of some universal set U. This set U will vary from problem to problem; sometimes we may take it to be the set of all integers, $\mathbb{Z}$, or the set of non-negative integers, $\mathbb{N}$, sometimes the defining attributes of an object, sometimes just an arbitrary collection of items used to construct an example.

First, we will meet some useful notation. If we want to specify the set of all the elements of the universe satisfying some condition P, we write

$$\{x \in U | P(x)\}$$

which is read 'The set of all objects x in the universe such that $P(x)$ is true', or 'The set of all objects x in the universe such that P is true for x'. This kind of notation is sometimes called **set-builder notation**.

Example

2.3

1. The set of all integers greater than -5 may be written as $\{x \in \mathbb{Z} | x > -5\}$.
2. The set $\{x \in \mathbb{Z} | x^2 > 16\}$ is the set of all integers whose square is greater than 16. Since this consists of those integers greater than 4 together with those integers less than -4, we could also write this set as $\{x \in \mathbb{Z} | (x > 4) \vee (x < -4)\}$.

If we have some sets, say A and B, in a given universe U, then we can construct new sets out of them in various useful ways, as follows:

1. **Union** $A \cup B = \{x \in U | x \in A \vee x \in B\}$
2. **Intersection** $A \cap B = \{x \in U | x \in A \wedge x \in B\}$

3. **Set difference** $A \setminus B = \{x \in U | x \in A \land x \notin B\}$
4. **Complement** $A' = U \setminus A = \{x \in U | x \notin A\}$

Example

2.4 Let the universe for the moment consist of the set $U = \{a, b, c, d, e, f, g\}$, and let $A = \{a, b, c, d\}$ and $B = \{c, d, e, f\}$. Then

1. $A \cup B = \{a, b, c, d, e, f\}$ 2. $A \cap B = \{c, d\}$ 3. $A \setminus B = \{a, b\}$
4. $B \setminus A = \{e, f\}$ 5. $A' = \{e, f, g\}$ 6. $B' = \{a, b, g\}$
7. $(A \cup B)' = \{g\}$ 8. $A' \cap B' = \{g\}$

If you examine the above examples and exercises, you will observe that some of the equalities are very similar to the algebraic laws we know for propositions. In fact, this is not too surprising when we consider how the set operations are defined. We can argue directly about unions and intersections of sets by arguing about the propositions that define them.

Worked example

2.5 Show that $(A \cup B)' = A' \cap B'$ for any two sets A and B.

Solution Whatever universe A and B lie in, we know that

$$\begin{aligned}(A \cup B)' &= \{x \in U | x \in A \lor x \in B\}' \\ &= \{x \in U | \neg(x \in A \lor x \in B)\} \\ &= \{x \in U | \neg(x \in A) \land \neg(x \in B)\} \\ &= \{x \in U | \neg(x \in A)\} \cap \{x \in U | \neg(x \in B)\} \\ &= A' \cap B'\end{aligned}$$

Self-Assessment Questions 2.2

Do you expect an equality involving sets corresponding to each one of the algebraic laws for propositions? Why?

Exercise 2.2

1. What numbers are in the set $\{x \in \mathbb{N} | x^2 < 100 \land x > 4\}$?
2. Express the set of numbers whose cube lies between -27 and 27 in set-builder notation.
3. If the universe U is $\{1, 2, 3, 4, 5, 6, 7, 8, 9, 10\}$, and $A = \{1, 3, 5, 7\}$, $B = \{4, 5, 6, 7, 8\}$ and $C = \{5, 6, 7, 8, 9, 10\}$, then

(a) Find $A \cup B$, $A \cup C$, $A \cap B$, A', C', $A \cap (B \cup C)$, $A \cup (B \cap C)$.
(b) Show that $A \cap (B \cup C) = (A \cap B) \cup (A \cap C)$, $(A \cap C)' = A' \cup C'$, $A \setminus (A \setminus B) = A \cap B$ and $A \setminus (B \setminus C) = A \cap (B' \cup C)$.

4. Let the universe be the set of attributes, $U = \{$four-legged, two-legged, hairy, feathered, scaly, warm-blooded, cold-blooded$\}$, and let lizards be described by the set

$$L = \{\text{four-legged, scaly, cold-blooded}\},$$

sauropods by the set

$$S = \{\text{four-legged, scaly, warm-blooded}\},$$

dogs by

$$D = \{\text{four-legged, hairy, warm-blooded}\},$$

and birds by

$$B = \{\text{two-legged, feathered, warm-blooded}\}.$$

(a) What set expressions correspond to those attributes common to (i) lizards and sauropods, (ii) sauropods and dogs, (iii) dogs and birds, (iv) sauropods, dogs and birds, and (v) all four types of animal?
(b) What do the set expressions (i) $S \cap B$ and (ii) $L \cap D$ correspond to?

5. (a) By a similar argument to that used in Worked example 2.5, show that for any three sets A, B and C, $A \cap (B \cup C) = (A \cap B) \cup (A \cap C)$.
(b) How can you express the universe and the empty set in set-builder notation?

2.3 Venn diagrams and membership tables

For simple problems involving few sets, it is often convenient to use diagrams to aid the understanding. Venn diagrams provide one method of doing this.

Venn diagrams

As is shown in Figure 2.1, we represent each set by a circle, and shade the region of interest. We can use such diagrams to represent expressions involving two or three sets; with more, it becomes difficult to include all the appropriate regions.

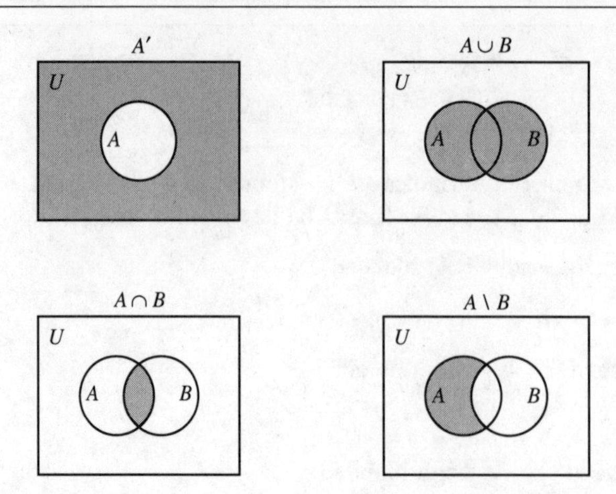

Figure 2.1

Worked example

2.6 Draw the regions $(A \cup B)'$ and $A' \cap B'$ on a Venn diagram, and check that they are the same.

Solution We first shade $A \cup B$ and then shade the complement differently to highlight it. Next, we shade A' and B' in another diagram, and consider the region that was shaded both times. As we see in Figure 2.2, the two are identical.

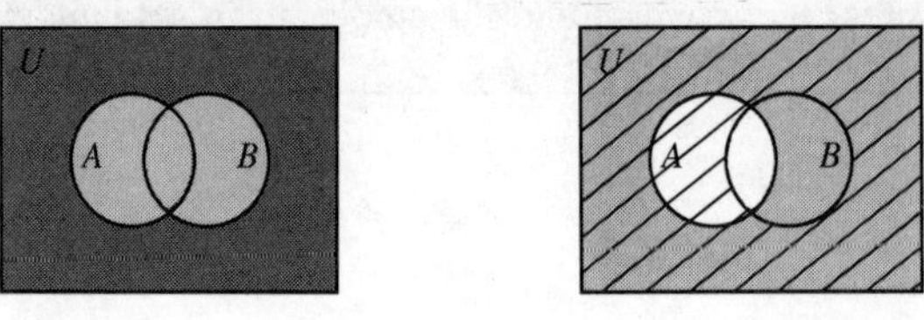

Figure 2.2

Venn diagrams, though limited in their uses by the fact that they are difficult to work with when more than three sets are considered, can motivate another approach. Let us consider the problem of deciding whether two expressions, involving combinations of three sets A, B and C, represent the same set. Now, by considering the Venn diagram in Figure 2.3, we see that there are eight regions that an element of the universe may lie in with respect to these three sets. It may lie in none of them, exactly one (three possibilities), exactly two of them (three possibilities), or in all three (one possibility). These eight regions and various unions of them are the only ones that can be specified by expressions involving A, B and C. So two expressions are equivalent in the sense that they specify equal sets just when the two expressions represent the same combination of these eight regions.

This hints at the following method. We list all the eight possible regions, and for each of the expressions we are interested in, we figure out which of the regions are included. If the same collection of these regions is involved in

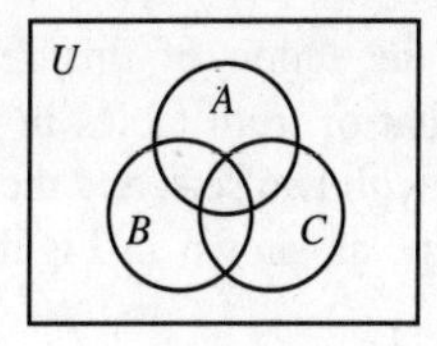

Figure 2.3

each, they represent the same set. This method is much more easily generalized to more than three sets, because it can be used in a tabular manner, without the need for diagrams. The version of it we will consider is known as using **membership tables**. There is a strong formal resemblance between membership tables and truth tables.

Membership tables

Armed with this insight gleaned from the consideration of the Venn diagram for three sets, let us see how to deal with expressions involving any number of sets. We want to know what all the possible regions are on a Venn diagram; and each region corresponds to those elements of the universe which are in the intersection of some collection of the sets in the expression, while being outside all the others. In other words, we can specify each region defined by some expression involving given sets by labelling each of the sets containing it with an I (for 'in'), and those sets not containing it with an O (for 'out'). We can therefore see that all the possible regions correspond to all the ways of assigning Is to some sets and Os to the others. We can then tabulate these in some systematic way.

So the rows of the table in Figure 2.4 describe all the regions in a Venn diagram for two sets, while those in the table in Figure 2.5 do the same job for a diagram for three sets.

A	B
I	I
I	O
O	I
O	O

Figure 2.4

A	B	C
I	I	I
I	I	O
I	O	I
I	O	O
O	I	I
O	I	O
O	O	I
O	O	O

Figure 2.5

And it is easy to see how to extend the pattern to deal with the case of more than three sets, although it is hard to draw the diagrams.

The next stage is to figure out just which regions lie in the set described by a given expression. We can do this by building up the result step by step. For example, in the set $A \cup (B \cap C)$, we first see which regions are in $B \cap C$, and then how many are in the union of this region with A. In fact, the technique is very similar to that of truth tables in propositional calculus. We enumerate all the possibilities with two sets, and then use these to construct the tables for arbitrary expressions, as shown in Figure 2.6. These tables tell us that

1. the regions in the complement of a set are precisely the ones not in the set.
2. the regions in the union of two sets are all those in at least one of the sets.
3. the regions in the intersection of two sets are precisely those in both sets.

Note that we do not need a table for $A \setminus B$, since $A \setminus B = A \cap B'$.

A	A'
I	O
O	I

A	B	$A \cup B$
I	I	I
I	O	I
O	I	I
O	O	O

A	B	$A \cap B$
I	I	I
I	O	O
O	I	O
O	O	O

Figure 2.6

We can use this to construct a membership table for any expression involving any number of sets; the rows of the table tell us which regions in the Venn diagram constitute the expression. We can see then that two expressions, E_1 and E_2, will represent the same set if they correspond to the same rows, and that $E_1 \subseteq E_2$ if all the regions in the set described by E_1 are in the set described by E_2.

Worked example

2.7 1. Show that the sets $A \cap (B \cup C)$ and $(A \cap B) \cup (A \cap C)$ are equal.
2. Show that $A \cap B \subseteq A$.

Solution 1. The tables for the two sets may be constructed together, as we see in Figure 2.7. As with the truth tables, the final row shows the order in which the columns are completed. Since columns 9 and 12 are identical, the sets are equal.

A	B	C	A	$\cap$	$(B$	$\cup$	$C)$	$(A$	$\cap$	$B)$	$\cup$	$(A$	$\cap$	$C)$
I	I	I	I	I	I	I	I	I	I	I	I	I	I	I
I	I	O	I	I	I	I	O	I	I	I	I	I	O	O
I	O	I	I	I	O	I	I	I	O	O	I	I	I	I
I	O	O	I	O	O	O	O	I	O	O	O	I	O	O
O	I	I	O	O	I	I	I	O	O	I	O	O	O	I
O	I	O	O	O	I	I	O	O	O	I	O	O	O	O
O	O	I	O	O	O	I	I	O	O	O	O	O	O	I
O	O	O	O	O	O	O	O	O	O	O	O	O	O	O
			1	9	2	8	3	4	10	5	12	6	11	7

Figure 2.7

2. Now, comparing the columns for A and $A \cap B$ in Figure 2.8, we see that only the region corresponding to the first row lies in $A \cap B$, but those corresponding to the first and second rows lie in A. Thus every region in $A \cap B$ is also in A, and so $A \cap B \subseteq A$.

Figure 2.8

A	B	A	$\cap$	B	A
I	I	I	I	I	I
I	O	I	O	O	I
O	I	O	O	I	O
O	O	O	O	O	O

Self-Assessment Questions 2.3

Explain how the regions of the Venn diagram correspond to rows of a membership table. Draw a Venn diagram for three sets, and label the regions with the number of the corresponding row in a membership table.

Exercise 2.3

1. Check the distributive law $A \cap (B \cup C) = (A \cap B) \cup (A \cap C)$ by means of a Venn diagram.
2. (a) Use a membership table to show that $A' \cup B' = (A \cap B)'$. Note: you will need to use a column for the complement of each of A, B and $A \cap B$.
   (b) Use a membership table to show that $A \cap (B \cap C) \subseteq A \cap (B \cup C)$.

2.4 Algebraic laws

It is worth comparing our current situation with that of the propositional calculus. As before, we now have an extremely methodical way of checking set expressions to see if they are actually equivalent, in the sense that they represent the same sets. However, also as before, this method suffers from the fact that the size of the tables grows very quickly as a function of the number of sets involved. Fortunately, and, yes, again as before, we can find a set of algebraic laws that will allow us to manipulate set expressions without the need to construct huge membership tables. Also, just as the membership tables look very similar to truth tables, so do the laws of set algebra look very similar to those for propositions. This type of similarity is called **isomorphism.**

First, however, we present the laws of set algebra. Let A, B and C be any sets in the universe U.

KEY POINT

1. Commutativity

$$A \cup B = B \cup A \qquad A \cap B = B \cap A$$

2. Associativity

$$A \cup (B \cup C) = (A \cup B) \cup C \qquad A \cap (B \cap C) = (A \cap B) \cap C$$

3. Distributivity

$$A \cup (B \cap C) = (A \cup B) \cap (A \cup C)$$
$$A \cap (B \cup C) = (A \cap B) \cup (A \cap C)$$

4. Idempotency

$$A \cup A = A \qquad A \cap A = A$$

5. de Morgan's Laws

$$(A \cup B)' = A' \cap B' \qquad (A \cap B)' = A' \cup B'$$

6. Properties of Complement

$$A'' = A \quad A \cup A' = U \quad A \cap A' = \emptyset \quad U' = \emptyset \quad \emptyset' = U$$

7. Properties of U and $\emptyset$

$$A \cup U = U \quad A \cap U = A \quad A \cup \emptyset = A \quad A \cap \emptyset = \emptyset$$

Each of these can be checked by using the definitions of $\cap$ and $\cup$ given above, or using membership tables. Recalling also that $A \setminus B$ is equivalent to $A \cap B'$, we can use these laws to make set algebraic expressions easier to understand.

Worked example

2.8 Show that

1. $A \setminus (A \setminus B) = A \cap B.$
2. $A = A \cap (A \cup B)$

Solution 1.

$$\begin{aligned} A \setminus (A \setminus B) &= A \cap (A \cap B')' \\ &= A \cap (A' \cup B'') && \text{(de Morgan)} \\ &= A \cap (A' \cup B) && \text{(complement)} \\ &= (A \cap A') \cup (A \cap B) && \text{(distributivity)} \\ &= \emptyset \cup (A \cap B) && \text{(complement)} \\ &= A \cap B && (U \text{ and } \emptyset) \end{aligned}$$

2. $A \cap (A \cup B) = (A \cup \emptyset) \cap (A \cup B)$ (U and $\emptyset$)
   $= A \cup (\emptyset \cap B)$ (distributivity)
   $= A \cup \emptyset$ (U and $\emptyset$)
   $= A$ (U and $\emptyset$)

Self-Assessment Question 2.4

Why are the laws of set algebra so similar to those for propositions?

Exercise 2.4

1. Prove a selection of the laws of set algebra using membership tables.
2. Show that $A = A \cup (A \cap B)$ using algebraic manipulation.
3. Prove each of the following identities using algebraic manipulation.
   (a) $A \setminus (B \cap C) = (A \setminus B) \cup (A \setminus C)$
   (b) $(A \cup B) \setminus (A \cap B) = (A \setminus B) \cup (B \setminus A)$

2.5 Counting

We will now consider the relationships between the numbers of elements contained in combinations of sets. This enables us to solve certain simple problems, and also lays a groundwork for later work on probability.

The number of elements of a set is also called its **cardinality**, and the cardinality of the set A is denoted $|A|$.

Worked example

2.9 What is the cardinality of the set $\{a, b, c, 1\}$?

Solution The set has 4 elements, so its cardinality is 4.

Now, what happens if we start with two sets, A and B? At first, it is a little tempting to think that $|A \cup B| = |A| + |B|$, but we can easily see that this isn't quite right, since if $A = \{a, b, c\}$ and $B = \{b, c, d\}$, we have $A \cup B = \{a, b, c, d\}$, so that $|A \cup B| = 4$, while $|A| + |B| = 3 + 3 = 6$. What has gone wrong is that we have counted the elements of $A \cap B$ both times, but each only occurs once in $A \cup B$. From this we see that

$$|A \cup B| = |A| + |B| - |A \cap B|$$

Worked example

2.10 The managerial requirements of a company call for two committees, one in charge of purchasing and one in charge of budgeting. The purchasing committee should have 27 members, the budgeting one should have 16. If no more than 8 people may be on both committees so as not to allow too much conflict of interest, and there are 30 people altogether who are qualified to be on one of these committees, can the committees be formed?

Solution We denote by P the purchasing committee, and by B the budgeting committee. Then we have $|P \cup B| = |P| + |B| - |P \cap B|$. But we know that $|P \cup B|$ cannot be more than 30, since there are only 30 people available. Also, $|P|$ is 27, $|B|$ is 16, and $|P \cap B|$ is at most 8. Thus $|P| + |B| - |P \cap B|$ is at least $27 + 16 - 8 = 35$. Thus the committees cannot be formed.

When there are only two sets involved, it is easy to solve such problems. With three or more, however, the situation becomes more complicated. For example, given three sets A, B and C, we have

$$|A \cup B \cup C| = |A| + |B| + |C| - |A \cap B| - |A \cap C| \\ - |B \cap C| + |A \cap B \cap C|$$

Although formulae such as this can easily be derived for more sets, they are not particularly useful in the context of the kind of problems we are considering.

We will consider a strategy for dealing with counting problems involving up to three sets, and use Venn diagrams as a tool. The general strategy for figuring out how many items lie in each of the various sets and their combinations – *i.e.* how many lie in each of the regions in a Venn diagram – is to begin in the middle and work out. In regions where the number of items is not given explicitly, simply use an unknown. Finally, all the information available should enable us to work out any unknowns we are interested in.

Worked example

2.11 Each member of a software development team of 16 is involved with at least one of systems analysis, coding, and maintenance. 10 worked on systems analysis, 6 on coding, and 9 on maintenance; 3 worked on systems analysis and coding, 3 on coding and maintenance, and 2 on all three. How many were involved in systems analysis and maintenance, but not coding?

Solution We draw a Venn diagram and fill in all we can, denoting by S the set of people involved in systems analysis, by C those involved in coding, and by M those involved in maintenance. So we fill in the triple intersection, and then the two known double intersections. We use an unknown, x, for the region we have no information on. Finally, the number of people in C tells us how many are involved in coding alone, the number in M tells us that those involved in coding alone amount to $6 - x$, while that in S tells us that those involved in systems analysis alone amount to $7 - x$. Adding all these numbers together, we should obtain 16, so that $6 - x + 1 + 2 + x + 2 + 1 + 7 - x = 16$, *i.e.* $19 - x = 16$, so that $x = 3$. But this is just the number that we want, so that 3 people were involved in systems analysis and maintenance, but not coding. The resulting diagram is shown in Figure 2.9.

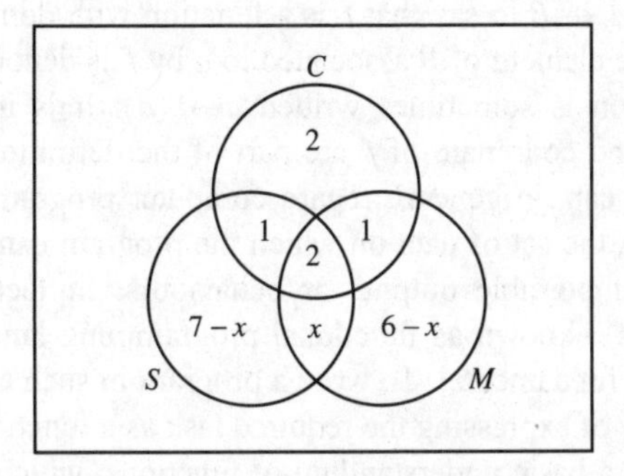

Figure 2.9

These simple counting methods that we have been looking at can be used to test the consistency of sets of requirements, as in the examples. They also provide an introduction to the kind of analysis involved in probability calculations, which we will consider later.

Self-Assessment Questions 2.5

Why can't we find the number of elements of the union of several sets just by adding the numbers in each set? When can we do this?

Exercise 2.5

1. (a) If a total of 27 people in a sample know at least one programming language, and of them 12 know PASCAL and 22 know COBOL, how many must know both?
   (b) There are 7 software packages available to do a task I am interested in, all of which can use modules in at least one of FORTRAN and C. If 5 can use modules in FORTRAN, and 4 can use modules in C, how many can use modules in both?
2. A team manager wishes to construct three overlapping teams, A, B and C. He has a pool of 40 to work with, and wants each team to contain 17 members. Furthermore, to ensure decent communication between teams, he wants 6 people to be in each of $A \cap B$, $A \cap C$, and $B \cap C$. Is this possible?

2.6 Functions

We will begin by considering the general properties and definitions involved with functions, and then go on to consider functions defined by formulae so that we can do some computations.

Functions on arbitrary sets

A **function** f is a rule associating to each element of some set A (called the **domain** of f) some element of another set B (called the **codomain** of f). We write $f : A \to B$ to say that f is a function with domain A and codomain B. If $a \in A$, the element of B associated to a by f is denoted $f(a)$. This last piece of information is sometimes written $a \mapsto f(a)$. It is important to note that the domain and codomain of f are part of the definition.

One can, in general, regard computer programs as functions, where the domain is the set of data on which the program can run, and the codomain is the set of possible outputs or behaviours. In fact, there are programming languages – known as functional programming languages – which take this notion as fundamental. To write a program in such a language, one has to find some way of expressing the required task as a function. We will concentrate on acquiring a basic understanding of functions, which will be applicable to any of the areas in which functional ideas are important.

Example

2.12 Let $A = \{1, 2, 3, 4\}$ and $B = \{a, b, c, d, e\}$. Then if f is defined by $f(1) = a$, $f(2) = c, f(3) = e, f(4) = a$ (or, alternatively, by $1 \mapsto a$, $2 \mapsto c$, $3 \mapsto e$, and $4 \mapsto a$) we have a function with domain A and codomain B.

This example serves to illustrate a few facts.

1. Every element of the domain must be sent to some element of the codomain.
2. Some elements of the codomain may fail to have any element of the domain sent to them.
3. No element of the domain is sent to more than one element of the codomain.
4. Elements of the codomain may have more than one element of the domain sent to them.

The subset of the codomain consisting of all those elements to which something is sent by f is called the **range** of f. We can use the set-builder notation to give another description of the range of f:

$$\text{range}(f) = \{f(a)|a \in A\}$$

We can also use a picture to represent a function, though this is only useful if there are few elements in the domain and codomain. This type of picture is sometimes called an **arrow diagram**. For example, the arrow diagram for Example 2.12 is shown in Figure 2.10. By convention, the domain is shown to the left and the codomain to the right of the diagram, and the arrows point from domain to codomain.

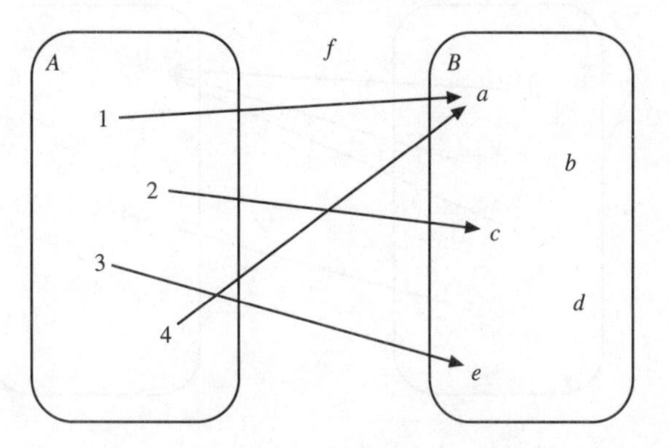

Figure 2.10

We observe that in the example above, the range is not all of the codomain, and some elements of the codomain have more than one element of the domain sent to them.

If the range is the whole of the domain, *i.e.* if no matter what element of the codomain you pick, there is some element of the domain which the function sends there, then the function is said to be **surjective**, or **onto**. In other words, in terms of the arrow diagram for the function, surjectivity means that each element of the codomain has an arrow pointing at it.

If no two elements of the domain are sent to the same element of the codomain, then the function is said to be **injective**, or **one-to-one**. In terms of the arrow diagram injectivity for a function means that no element of the codomain has more than one arrow pointing at it.

If a function is both injective and surjective, it is also called **bijective**, or **invertible**, since in this case we can find a function that takes each element of the codomain back to the element of the domain it started as; in other words, we can invert, or undo, the process that the function carries out. In terms of the arrow diagram, each element of the codomain has exactly one arrow pointing at it, so that we can unambiguously say what element of the domain the function sends to each element of the codomain.

A mental picture which may be helpful is to think of a function as being a rule that assigns a label to each object in some set; the set of objects is the domain, the set of possible labels the codomain. Then the function is surjective if all labels are used up, and injective if no two items get the same label.

Worked example

2.13 Let $A = \{1, 2, 3, 4\}$, $B = \{a, b, c\}$ and $C = \{x, y, z, w\}$. Draw the arrow diagram for each of the following, and say whether it is bijective, injective, surjective, or none of these.

1. $f : A \to B$, with $f(1) = a, f(2) = a, f(3) = a, f(4) = b$
2. $f : B \to A$, with $f(a) = 2, f(b) = 1, f(c) = 4$
3. $f : C \to A$, with $f(x) = 4, f(y) = 2, f(z) = 1, f(w) = 3$

Solution 1. Nothing is mapped to c, and $f(1) = f(2)$, so this function is neither injective nor surjective; see Figure 2.11.

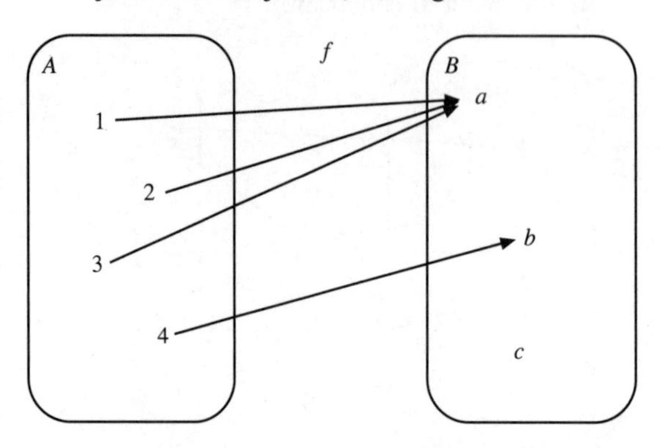

Figure 2.11

2. Nothing is sent to 3, so the function is not surjective; however, no two items are sent to the same result, so the function is injective. See Figure 2.12.
3. No two elements are sent to the same result, and all the codomain gets used up, so the function is bijective. The arrow diagram for this function is shown in Figure 2.13.

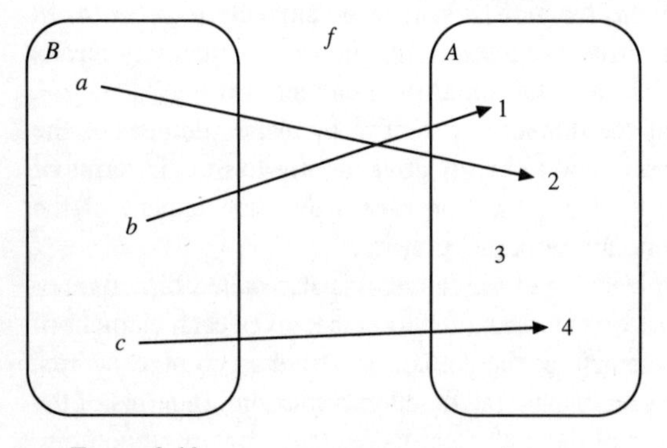

Figure 2.12

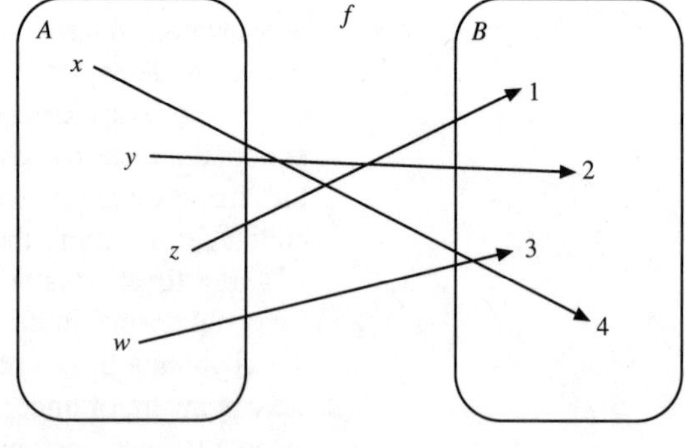

Figure 2.13

Composition of functions

If we have two functions, $f : A \to B$ and $g : B \to C$, then we can construct a new function $g \circ f : A \to C$ simply by feeding each element of A to f, and feeding the result to g. The corresponding diagram is that of Figure 2.14 and we write $g \circ f(a) = g(f(a))$.

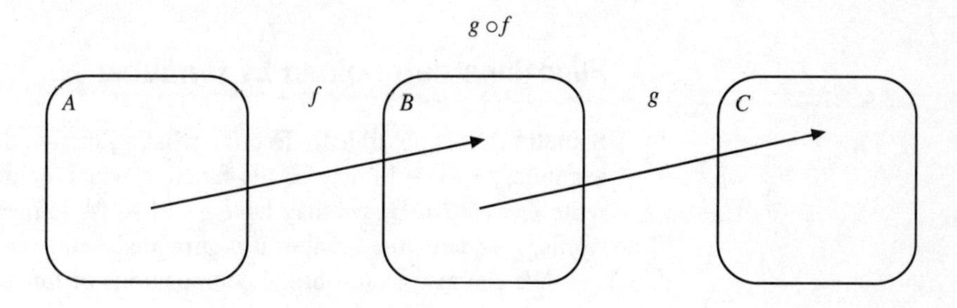

Figure 2.14

In terms of programs, we can think of running two programs; the output from the first is the input to the second.

Worked example

2.14 What is $g \circ f$ if $f : \{1,2,3\} \to \{x,y\}$ by $f(1) = x$, $f(2) = y$, $f(3) = x$, and $g : \{x,y\} \to \{\alpha, \beta\}$ by $g(x) = \beta$, $g(y) = \alpha$.

Solution We first note that $g \circ f : \{1,2,3\} \to \{\alpha, \beta\}$. Next, $g \circ f(1) = g(f(1)) = g(x) = \beta$, $g \circ f(2) = g(f(2)) = g(y) = \alpha$ and $g \circ f(3) = g(f(3)) = g(x) = \beta$.

Inverse functions

If we have a function $f : A \to B$, then for any element b of B, the set $f^{-1}(b)$ is defined as $\{a \in A | f(a) = b\}$. For a general function, this set may have no elements at all, or it may have several. We are especially interested in the case where it has exactly one element for each $b \in B$, since that means that we can undo the action of f.

We denote by 1_A the **identity function** on A, namely the function defined by $1_A(a) = a$ for any $a \in A$. If $f : A \to B$ is bijective, then there is a function $g : B \to A$ such that $f \circ g = 1_B$ and $g \circ f = 1_A$. We also write $f = g^{-1}$ and $g = f^{-1}$, and say that f is the inverse of g and that g is the inverse of f. In other words, the inverse of a function undoes whatever the function does. If $f(a) = b$, then $f^{-1}(b) = a$.

Worked example

2.15 Show that $f : \{1,2\} \to \{a,b\}$ defined by $f(1) = b$, $f(2) = a$ is invertible, and find f^{-1}.

Solution f is injective and surjective. Since $f(2) = a, f^{-1}(a) = 2$, and since $f(1) = b$, $f^{-1}(b) = 1$, so f^{-1} is given by $f^{-1}(a) = 2, f^{-1}(b) = 1$.

Functions determined by formulae

Sometimes it is possible to describe what a function does using a formula. For example, $f : \mathbb{N} \to \mathbb{N}$ may be the function which doubles any number, and we write $f(n) = 2n$. Or we may have $g : \mathbb{N} \to \mathbb{N}$ defined by $g(n) = n^2$, in other words, g squares the number it is provided with.

We can also work out the compositions of functions defined in this way, using the definitions from above.

Worked example

2.16 If $f, g : \mathbb{N} \to \mathbb{N}$ are defined by $f(n) = 3n$ and $g(n) = n^2$, then what are $f \circ g$ and $g \circ f$?

Solution By definition, $f \circ g(n) = f(g(n)) = f(n^2) = 3n^2$. Similarly, $g \circ f(n) = g(f(n)) = g(3n) = (3n)^2 = 9n^2$.

Self-Assessment Questions 2.6

If you think of a program as a function, what does it mean to say that the function is invertible? What does the inverse function calculate in this case?

Exercise 2.6

1. Let $f : \{x, y, z\} \to \{2, 4, 6\}$ be defined by $f(x) = 4$, $f(y) = 2$, $f(z) = 4$. What is the range of f?
2. Let $f : \{x, y, z\} \to \{2, 4, 6\}$ be defined by $f(x) = 4$, $f(y) = 2$, $f(z) = 4$. What does the arrow diagram for f look like?
3. Draw the arrow diagrams for each of the following, and classify them as bijective, injective, surjective, or none of these.

   (a) $f : \{a, b, c\} \to \{a, b, c\}$ given by $f(a) = c, f(b) = a, f(c) = c$.
   (b) $f : \{a, b, c\} \to \{a, b, c\}$ given by $f(a) = c, f(b) = a, f(c) = b$.
   (c) $f : \{a, b\} \to \{a, b, c\}$ given by $f(a) = c, f(b) = a$.
4. If $f : \{x, y, z, w\} \to \{2, 4, 6\}$ where $f(x) = 4, f(y) = 6, f(z) = 4$ and $f(w) = 2$, and $g : \{2, 4, 6\} \to \{\alpha, \omega\}$ where $g(2) = \alpha$, $g(4) = \alpha$, and $g(6) = \omega$, then find $g \circ f$ and draw the arrow diagram.
5. Which of the following functions has an inverse? In each case where the inverse exists, say what it is. If an inverse does not exist, say why not.

   (a) $f : \{a, b, c\} \to \{a, b, c\}$ given by $f(a) = c, f(b) = a, f(c) = a$

(b) $f : \{a, b, c\} \to \{a, b, c\}$ given by $f(a) = c, f(b) = a, f(c) = b$

(c) $f : \{a, b, c\} \to \{a, b, c\}$ given by $f(a) = c, f(b) = a$

6. We denote by $\mathbb{R}$ the set of all real numbers, i.e. we include fractions and square roots, and all the rest.

   (a) If $f : \mathbb{R} \setminus \{1\} \to \mathbb{R}$ is given by $f(x) = (1+x)/(1-x)$, what is $f \circ f$?

   (b) If $f : \mathbb{N} \to \mathbb{R}$ is given by $f(n) = \sin n$, and $g : \mathbb{R} \to \mathbb{R}$ is given by $g(x) = x^2$, what is $g \circ f$?

   (c) Is $f : \mathbb{R} \to \mathbb{R}$ given by $f(x) = x^2$ invertible? If so, what is the inverse?

   (d) Is $f : \mathbb{R} \to \mathbb{R}$ given by $f(x) = x^3$ invertible? If so, what is the inverse?

2.7 Relations

There is another way in which we could think of a function. Rather than explicitly thinking of it as a rule associating elements of the domain to elements of the codomain, we could think of it as a list of pairs, where the first item in the pair is an element of the domain, and the second is what the first gets sent to in the codomain. In other words, we think of the function as specified by the list of all $(a, f(a))$ pairs for a in the domain of f.

To take this approach, we define the **Cartesian product** of two sets, A and B, by

$$A \times B = \{(a, b) | a \in A \land b \in B\}$$

Worked example

2.17 What is the product $\{1, 2\} \times \{a, b, c\}$?

Solution The product is given by $\{(1, a), (1, b), (1, c), (2, a), (2, b), (2, c)\}$.

With this definition as part of our mental toolkit, we see that a function with domain A and range B may be specified by a subset of the Cartesian product $A \times B$. In fact, since the function is completely specified by the subset, and *vice versa*, we can go so far as to say that the function is the subset. However, in order to reduce the confusion, we will denote the subset of $A \times B$ corresponding to f by R_f. However, a function defines a special kind of subset. There must be at least one element of the subset for each element of the range, and there can be no more than one. We can also see that a function being surjective corresponds to there being at least one element of the subset for each element of the range, and a function being injective corresponds to there being at most one element of the subset for each element of the range.

We can now consider subsets of $A \times B$ that fail to satisfy the conditions for a function. If we do this, we get what is called a **relation** from A to B. It is like a function, except that a single element of A may be associated with more than one element of B, and some may be associated with no element of B.

Example

2.18 Let $A = \{1, 2, 3\}$ and $B = \{a, b\}$. Then $R_1 = \{(1, a), (2, a), (3, b)\}$ corresponds to a function, and $R_2 = \{(2, a), (2, b)\}$ does not. Both, however, are relations.

Data are often stored in computers in such a way that it is possible to get from one item to another by looking for common properties; this provides the motivating idea for relational databases. With just one additional idea, we can begin to see how these set-algebraic notions can tie in with relational databases. Given two sets, A and B, we can form their Cartesian product, $A \times B$. There are two naturally defined functions here, called the **projection** functions. They are $p_1 : A \times B \to A$, and $p_2 : A \times B \to B$, defined by $p_1(a, b) = a$ and $p_2(a, b) = b$; in other words, p_1 gives us the first item in the pair, and p_2 gives us the second. How this can be used is possibly best seen by means of an example.

Worked example

2.19 Consider the following list of books with their authors:

1. *Emma* by Jane Austen
2. *The Idiot* by F Dostoyevsky
3. *Winnie the Pooh* by AA Milne
4. *Pride and Prejudice* by Jane Austen
5. *Crime and Punishment* by F Dostoyevsky
6. *Mansfield Park* by Jane Austen
7. *The Idiot* by N Oman

Construct a relation that contains this information, and, using this relation and the projection functions, show how to express the set of all books by the same author as *Mansfield Park*.

Solution We have the set $T = \{$*Emma, The Idiot, Winnie the Pooh, Pride and Prejudice, Crime and Punishment, Mansfield Park*$\}$ of titles, and $A = \{$Jane Austen, F Dostoyevsky, AA Milne, N Oman$\}$ of authors. The Cartesian product $T \times A$ of these is quite large, but our data may be encapsulated as the subset $\{$(*Emma*, Jane Austen), (*The Idiot*, F Dostoyevsky), (*Winnie the Pooh*, AA Milne), (*Pride and Prejudice*, Jane Austen), (*Crime and Punishment*, F Dostoyevsky), (*Mansfield Park*, Jane Austen), (*The Idiot*, N Oman)$\}$.

Now,

$p_1^{-1}(\textit{Mansfield Park})$

gives us all the pairs in the relation where *Mansfield Park* is the first element.

$$p_2(p_1^{-1}(\textit{Mansfield Park}))$$

gives us all the authors who have written books called *Mansfield Park.*

$$p_2^{-1}(p_2(p_1^{-1}(\textit{Mansfield Park})))$$

gives all pairs consisting of books by these authors, and so

$$p_1(p_2^{-1}(p_2(p_1^{-1}(\textit{Mansfield Park}))))$$

is the list of all titles of books by anybody who has written a book called *Mansfield Park.*

Note: we have used the notation $f(X)$ as a shorthand for $\{f(x) : x \in X\}$ in this example. It is a useful item of notation, and well worth remembering.

This provides just a small taste of how set-theoretic ideas may be used to represent and manipulate the relationships between items of data. The theory underlying relational databases is based on such ideas.

Self-Assessment Question 2.7

Why is it important to use relations rather than just functions for modelling this kind of relationship?

Exercise 2.7

In Worked example 2.19, how would you express the set of all people who have written a book with the same title as anything by F Dostoyevsky?

Test and Assignment Exercises 2

1. Let $U = \{1, 2, 3, 4, 5, 6, 7, 8, 9\}$ and consider the subsets A, B and C defined by $A = \{1, 3, 5, 7, 9\}$, $B = \{2, 4, 6, 8\}$ and $C = \{1, 2, 5, 6, 8, 9\}$.
   (a) Find each of the sets $A \cup C$, $B \cap C$, A', $C \setminus B$, $B \setminus C$, $A \cap (B \cup C)$, and $(A \cap B) \cup (A \cap C)$.
   (b) Draw each of these sets on a Venn diagram.
   (c) Is $\{1\}$ an element of U?
2. Let U be a universal set, with subsets A, B and C.
   (a) Draw Venn diagrams for each of $A \cup B$, $B \setminus A'$, $C \setminus B$, $A \cap B'$, $A \cup (B \cup C)$, $A \cup (B \cap C)$, $(A \cap B) \cap C$, $(A \cup B) \cap (A \cup C)$.
   (b) For those pairs where the Venn diagrams give the same area, prove equality using membership tables.

3. Prove the identities

   (*i*) $A \cup (A \cap B) = A$
   (*ii*) $(A \cap B \cap C) \cup (A' \cap C) \cup (B' \cap C) = C$
   (*iii*) $(A \cap B') \cup (B \cap A') = (A \cup B) \cap (A \cap B)'$
   (*iv*) $A \setminus (B \setminus (C \setminus D)) = (A \setminus B) \cup ((A \cap C) \setminus D)$
   (*v*) $(A \cup B) \setminus (A \cap B) = (A \setminus B) \cup (B \setminus A)$
   (*vi*) $(C \cup (A \cap B)) \setminus (A \cap C) = (C \setminus A) \cup ((A \cap B) \setminus C)$

   (a) by drawing Venn diagrams (when there are no more than three sets involved).
   (b) using membership tables.
   (c) using the laws of set algebra.

4. Using membership tables,
   (a) prove that for any sets A and B, $A \subseteq A \cup B$
   (b) show that $A \subseteq B$ if and only if $A = A \cap B$.

5. (a) If $(A \cup C) = (B \cup C)$, does it follow that $A = B$? If it does, prove it; if not, find a counter-example.
   (b) Same question as above, but with $\cap$ instead of $\cup$.
   (c) Show that if $A \cup C = B \cup C$ and $A \cap C = B \cap C$ for some set C, then $A = B$.

6. Define $|$ by $A|B = \overline{A} \cup \overline{B}$. Find expressions involving only A, B, and $|$ that have the same membership tables as $\overline{A}$, $A \cap B$, and $A \cup B$.

7. In a group of 70 students, each of whom studied at least one of the subjects economics, history and geography, 53 studied economics, 3 studied only history, 4 only geography, 35 economics and history and 34 history and geography. How many studied both economics and history but not geography?

8. In a survey of 1000 households, washing machines, vacuum cleaners and refrigerators were counted. Each house had at least one of these appliances. 400 had no refrigerator, 380 had no vacuum cleaner and 542 no washing machine. 294 had both a vacuum cleaner and washing machine, 277 both a refrigerator and a vacuum cleaner, 190 both a refrigerator and a washing machine. How many households had all three appliances? How many had only a vacuum cleaner?

9. Each person in a group of 150 can be described as one or more of tall, fat and unwashed. If 66 are tall, 37 are fat only, 45 are unwashed only, 26 are fat and tall and 11 are fat and unwashed, how many are tall and fat but not unwashed?

10. Let $A = \{1,2,3,4\}$, $B = \{a,b,c\}$ and $C = \{x,y,z,w\}$. What are $A \times B$ and $A \times C$? For each of the following relations $R \subset A \times B$ or $R \subset A \times C$ state whether R corresponds to a function. When it does, draw the arrow diagram for the function. Also, state whether each function is injective, surjective, bijective, or none of these.

    (a) $R = \{(1,a),(2,a),(1,b)\} \subset A \times B$
    (b) $R = \{(1,a),(2,c),(3,a)\} \subset A \times B$
    (c) $R = \{(1,a),(2,c),(3,a),(4,b)\} \subset A \times B$
    (d) $R = \{(1,a),(2,c),(3,a),(4,b),(2,a)\} \subset A \times B$

(e) $R = \{(1,a),(2,c),(3,a),(4,c)\} \subset A \times B$
(f) $R = \{(1,x),(2,y),(3,x),(4,z)\} \subset A \times C$
(g) $R = \{(1,x),(2,z),(3,w),(4,y)\} \subset A \times C$

11. One of the following relations gives an invertible function from the set $\{a,b,c\}$ to the set $\{x,y,z\}$. Find out which it is, and also give the arrow diagram for it; then give the arrow diagram and the list of ordered pairs for the inverse function.
(a) $\{(a,x),(c,z),(b,y)\}$ (b) $\{(a,z),(b,z),(c,x)\}$
(c) $\{(a,x),(b,y),(c,z),(a,z)\}$

12. Let A be the set $\{1,2,3,4,5\}$, B be $\{-2,-1,0,1,2\}$ and let C be $\{0,1,4\}$. If $f : A \to B$ is given by $f(n) = n-3$, and $g : B \to C$ is given by $g(m) = m^2$, then find a formula for $g \circ f : A \to C$, and also draw the arrow diagrams, and give the set of ordered pairs corresponding to each of f, g and $g \circ f$.

13. Bob, Carol, Ted and Alice are programmers who work together in various combinations on projects. At the moment, they are working on six projects, which we will call P1 to P6. The list of people working on each project is:
(a) P1: Bob, Ted and Carol. (b) P2: Alice. (c) P3: Bob and Carol. (d) P4: Bob and Alice. (e) P5: Bob and Ted. (f) P6: Bob, Carol, Ted and Alice.
If we call the set of projects P and the set of programmers R, find a relation (*i.e.* a subset of $P \times R$) that encapsulates the above information. Using the projection mappings and their inverses, find expressions for

(a) The programmers working on P3. (b) The projects that Ted is working on. (c) The projects worked on by all four programmers. (d) All projects worked on by anybody working on P5.

2.8 Further reading

H T Combe, *Sets and Symbolic Logic* (Ginn, 1970)
This provides further reading on the material covered in Chapters 1 and 2.
H F Korth and A Silverschutz, *Database System Concepts* (McGraw-Hill, 1991)
A computer science textbook which explains relational databases, and also various other approaches to the manipulation of data.
P R Halmos, *Naive Set Theory* (Van Nostrand, 1990)
A careful development of set theory which also introduces the axiomatic approach.
S Lipschutz, *Set Theory* (McGraw-Hill, 1964)
A text relying heavily on worked examples, also providing further reading on the material of this chapter.
M McFadden, J W Moore and W I Smith, *Sets, Relations and Functions* (McGraw-Hill, 1963)
Another text providing a lengthier exposition of the material of this chapter.

3 Predicate calculus

Objectives

When you have mastered this chapter you should be able to

- understand the concept of a **predicate** as a truth valued function
- understand the concept of a **quantifier**
- translate between statements in natural language and in quantified predicate form
- use the rules and relations of predicate calculus to reason about the validity of statements represented as quantified predicates

In this chapter we will meet an extension of the propositional calculus known as the **predicate calculus**. We shall see how the predicate calculus can be used to represent and to reason about quite complicated statements. The predicate calculus is an important tool in computer science and information technology. It is used in formal program development methods and forms a basis for methods used in reasoning about relational databases.

3.1 Predicates

In chapter 1 we met statements called **propositions** that were characterized by the fact that they were clearly either *true* or *false*. In this section we will meet functions called predicates that give values that are sometimes *true* and sometimes *false*.

Predicate functions

In chapter 1 we saw how the Propositional Calculus could be used to reason about statements that had a definite truth value. An example of such a statement or proposition was 'The programming team is happy with the specification'. Such a proposition, dealing with just one programming team and one specification, is inadequate if we wish to reason about a variety of programming teams and specifications. A way of resolving this restriction is to create a function IsHappyWith(s, t), which takes two values, 't' from the set of programming teams and 's' from the set of specifications, and returns a value from the set $\{true, \; false\}$. So the function IsHappyWith(spec2, team1) will return a value of *true* if the programming team 'team1' is happy with the specification 'spec2' or a value of *false* if they are not. Similarly the proposition 'My program compiles' may be made general by introducing the function Compiles(p), which takes a value 'p' from the set of program identifiers and returns a truth value which indicates whether the identified program compiles or not. The functions IsHappyWith(s, t) and Compiles(p) are examples of a class of **functions** known as **predicates**.

Worked example

3.1 Create predicates that can be used to represent the following propositions.

(a) L N Tolstoy wrote 'War and Peace'. (b) 7 is a prime number.
(c) Mount Everest is higher than the Eiger.

Solution (a) Introduce a predicate Wrote(t, a) which takes two variables 'a' from the set of authors and 't' from the set of texts and returns the value *true* if the author 'a' wrote text 't' or *false* otherwise. We can now write the given proposition as:

$$\text{Wrote}(\text{WarandPeace}, \text{LNTolstoy})$$

(b) Introduce a predicate IsPrime(n) which takes a number n from the set of natural numbers and returns the value *true* if n is a prime number and *false* otherwise. The given proposition can now be written as:

$$\text{IsPrime}(7)$$

(c) Introduce the predicate IsHigherThan(M1,M2), which takes the names of two mountains, represented by the variables M1 and M2, and returns the value true if the mountain represented by M2 is higher than the mountain represented by M1. We can now write:

$$\text{IsHigherThan}(\text{Eiger}, \text{Everest})$$

Relational operators are predicates

From your experience of algebra you will be familiar with expressions of the form $x < y$ or $x = 2$. In these expressions x and y are variables that represent values from some set of numbers, perhaps the naturals $\mathbb{N}$ or the integers $\mathbb{Z}$.

When x and y are assigned particular numeric values the two expressions, $x < y$ and $x = 2$, will have particular truth values. For example if x is assigned the value -3 and y is assigned the value 6 then $x < y$ becomes $-3 < 6$, which is true, and $x = 2$ becomes $-3 = 2$, which is false. The symbols $=$ and $<$ are examples of relational operators as are $>$, $\leq$, $\geq$ and $\neq$. Since truth values are the result of evaluating expressions that contain relational operators, these operators are examples of predicates.

Worked example

3.2 Devise a predicate for which the proposition, '27 is 3 cubed', is a particular instance.

Solution Using variables x and n to represent values from the set $\mathbb{Z}$ of integers we can write:

$$x = n^3$$

Assigning actual values gives:

$$27 = 3^3$$

Definition of a predicate

Having considered some particular examples we are now in the position to provide a definition of what is meant by a predicate.

KEY POINT

Given the sets $S_1, S_2, \ldots, S_n$ and the set $\{true, \ false\}$, which we will represent as $\mathbb{B}$, we can define a predicate as any function P which is of the form:

$$P : S_1 \times S_2 \times \cdots \times S_n \to \mathbb{B} \text{ with } n \geq 1$$

Self-Assessment Question 3.1

Do you think that the symbol $\in$, which you met in the chapter on sets, represents a predicate? Justify your answer.

Exercise 3.1

1. Write the following propositions as instances of a predicate:

   (a) Henry VIII had six wives.
   (b) 4 is a factor of 16.
   (c) Lead is denser than iron.

2. Devise a predicate for which the following propositions are particular instances:

   (a) 4 squared is greater than twice 4.
   (b) 5 multiplied by 4 is greater than 5 plus four.
   (c) The square root of 9 is less than or equal to 9.

3.2 Combining predicates

Since predicates evaluate to truth values they may be combined together using the logical connectives that were introduced in the chapter on propositional calculus. The result of this combination of predicates and logical connectives is called the **predicate calculus**.

Worked examples

3.3 Represent the following statements using predicate calculus.

(a) If Jim is employed as a programmer and has less than three years experience then he will be working as member of team3 and will be managed by Sarah.
(b) The fact that Sally is the leader of team2 means that she is either a systems analyst with expertise in object oriented design (OOD) or she is a programmer with expertise in C++ and has more than four years' experience.

Solution With the following sets:

E set of employees

PT set of project teams

J set of job titles

S set of areas of expertise

we can introduce the following functions:

Experience: $E \to \mathbb{N}$
Indicates the number of years experience that a given employee has.

IsIn: $PT \times E \to \mathbb{B}$
Indicates whether a given employee is part of a given project team.

EmployedAs: $J \times E \to \mathbb{B}$
Indicates whether a given employee has a given job title.

Leader: $PT \times E \rightarrow \mathbb{B}$

Indicates whether a given employee leads a specified project team.

ExpertIn: $S \times E \rightarrow \mathbb{B}$

Indicates whether a given employee is an expert in the specified field.

Manages: $E \times E \rightarrow \mathbb{B}$

Indicates whether the first given employee is managed by the second.

We can now write the given statements using predicates and logical connectives. For part (a) we get:

$$\text{EmployedAs(programmer, Jim)} \wedge \text{Experience(Jim)} < 3 \Rightarrow \\ \text{IsIn(team3, Jim)} \wedge \text{Manages(Jim, Sarah)}$$

and for part (b) we get:

$$\text{Leader(team2, Sally)} \Rightarrow (\text{EmployedAs(systems analyst, Sally)} \wedge \\ \text{ExpertIn(OOD, Sally)}) \vee (\text{EmployedAs(programmer, Sally)} \wedge \\ \text{ExpertIn(C++, Sally)} \wedge \text{Experience(Sally)} > 4)$$

3.4 If S is the set of all sets from some universe U and if E is the set of all elements from the same universe use the following predicates:

$$\text{Equal}: S \times S \rightarrow \mathbb{B}$$
$$\text{SubsetOf}: S \times S \rightarrow \mathbb{B}$$
$$\text{IsIn}: S \times E \rightarrow \mathbb{B}$$

to represent the following statements.

(a) If the set S_1 is a subset of the set S_2 and the set S_2 is a subset of the set S_1 then both sets are equal.

(b) If S_1 is a subset of S_2 and e is an element of S_1 then e is also an element of S_2.

(c) e is an element of the intersection of S_1 and S_2 if and only if e is an element of S_1 and also an element of S_2.

Solution (a) In this solution we need to express the fact that if the set S_1 is a subset of the set S_2 and if the set S_2 is a subset of the set S_1 then the combination of these conditions is equivalent to the two sets being equal. This gives rise to the expression:

$$\text{SubsetOf}(S_2, S_1) \wedge \text{SubsetOf}(S_1, S_2) \Leftrightarrow \text{Equal}(S_1, S_2)$$

(b) In this solution we need to express the fact that set S_1 is a subset of the set S_2 and in consequence if set S_1 contains some element e then this element is also contained in set S_2. This gives rise to the expression:

$$\text{SubsetOf}(S_2, S_1) \wedge \text{IsIn}(S_1, e) \Rightarrow \text{IsIn}(S_2, e)$$

(c) In this solution we need to express the fact that if some element e is in the set produced by the intersection of S_1 and S_2 then this element must be in both S_1 and S_2. This gives rise to the expression:

$$\text{IsIn}(S_1 \cap S_2, e) \Leftrightarrow (\text{IsIn}(S_1, e) \wedge \text{IsIn}(S_2, e))$$

Exercise 3.2

1. Using the sets and functions from Example 3.3 express as a predicate the fact that if Sarah is Paul's manager then she either has more years' experience than Paul or is an expert in OOD and C++.
2. Using the sets and predicates from Example 3.4 express as a predicate the fact that if an element appears in the intersection of two sets then it is in the union of the two sets and also in each set.

3.3 Quantified predicates

In the previous section we saw how a combination of predicates and logical connectives may be used to represent statements about particular elements from given sets. For example we made statements about Jim from the set of employees and about 27 from the set $\mathbb{N}$ of natural numbers. What we cannot do at the moment is to represent statements that contain some aspect of quantification. Examples of such statements are: 'There are no team leaders that have less than four years experience' and 'There is only one even prime number'. In these two examples quantification manifests itself in the phrases: 'There are no...' and 'There is only one...'. In order to deal with statements of this type we have to introduce quantifiers.

The existential quantifier

Consider the statement 'There is at least one employee who is an expert in C++'. Now if there are n employees in the company, then using the notation from Example 3.3 above we know that the set E has n elements so we can write $E = \{e_1, e_2, \ldots, e_n\}$ and we can represent the above statement by the disjunction:

$$\text{ExpertIn}(\text{C++},\ e_1) \vee \text{ExpertIn}(\text{C++},\ e_2) \vee \cdots \vee$$
$$\text{ExpertIn}(\text{C++},\ e_n)$$

By introducing the symbol $\exists$ which is known as the **existential quantifier** we may write the above disjunction more compactly as:

$$\exists e \in E : \text{ExpertIn}(\text{C++},\ e)$$

The symbol $\exists$ may be expressed in English by such phrases as 'there exists' or 'there is at least one'. The colon ':' can be expressed as 'such that' or as 'that is'. So the predicate $\exists e \in E : \text{ExpertIn}(\text{C++},\ e)$ may be expressed as:

'There exists an employee "e", such that "e" is an expert in C++'. Or somewhat more elegantly as 'There is an employee who is an expert in C++'.

KEY POINT

> The existential quantifier is used when we wish to express the fact that for **at least one** value of the quantified variable the associated predicate becomes true.

Let us consider two more examples of the use of the existential quantifier. If we have the set $X = \{2, 3, 4, 5\}$ then we can express the fact that this set contains at least one prime number by the following disjunction:

$$\text{IsPrime}(2) \vee \text{IsPrime}(3) \vee \text{IsPrime}(4) \vee \text{IsPrime}(5)$$

We may abbreviate this disjunction to:

$$\exists x \in X : \text{IsPrime}(x)$$

Now consider the task of representing as a predicate the fact that there is a natural number which is both even and prime. Since the set of natural numbers contains an infinite number of members the thought of expressing the above fact in the form of a disjunction is somewhat daunting. The only practical way of creating the appropriate predicate is to use the existential quantifier. We can write:

$$\exists x \in \mathbb{N}\text{: IsPrime}(x) \wedge \text{IsEven}(x)$$

Worked examples

3.5 Express as a predicate the fact that there is a natural number whose square is 64.

Solution

$$\exists x \in \mathbb{N}\text{: } x^2 = 64$$

This predicate may be read as: 'There exists a natural number x such that x squared is 64'.

3.6 Using the sets and functions from Worked example 3.3 express as a predicate the fact that at least one employee has ten or more years' experience.

Solution

$$\exists e \in E : \text{Experience}(e) \geq 10$$

This predicate may be read as: 'There exists an employee identified by the letter e who when used as an argument to the function "Experience" gives a value of 10 or more'.

3.7 Since quantified predicates also evaluate to truth values they may be combined using logical connectives. Again using Worked example 3.3 express the following statements as predicates:

(a) If Bob is the leader of team4 then at least one of the members of the team will be an expert in OOD but will have less than three years' experience.
(b) There is no member of team1 who has less than four years' experience.

Solution (a) Leader(team4,Bob) $\Rightarrow \exists e \in E$: (IsIn(team4,$e$) $\wedge$ ExpertIn(OOD,e)$\wedge$ Experience(e) < 3)

In this solution the predicate on the left of the implication symbol asserts that Bob is the leader of team4. To the right of the implication symbol we have the consequence which consists of the conjunction of three existentially quantified predicates. The first predicate tells us that there is an employee who is in team4, the second that this employee is an expert in OOD and the third that the employee has more than three years' experience.

(b) $\neg\exists e \in E : (\text{IsIn}(\text{team1}, e) \wedge \text{Experience}(e) < 4)$

This solution employs the negation of an existential quantifier to state that there does **not** exist an employee with the characteristics described by the quantified predicates.

The universal quantifier

Consider the statement that 'Every employee has at least two years experience'. If we again use the sets and functions from Example 3.3 and state that the company has n employees we can again write $E = \{e_1, e_2, \ldots, e_n\}$. Consequently the above predicate may be expressed as:

$$\text{Experience}(e_1) \geq 2 \wedge \text{Experience}(e_2) \geq 2 \wedge \cdots \wedge \text{Experience}(e_n) \geq 2$$

By introducing the symbol $\forall$, which is known as the **universal quantifier**, we may abbreviate the above predicate to:

$$\forall e \in E : \text{Experience}(e) \geq 2$$

The symbol $\forall$ may be expressed in English as 'for every' or 'for all' or just simply as 'all'. So the above predicate simply states that 'All employees have two or more years experience'. Note that when expressing a universally quantified expression in English the colon : is not pronounced.

KEY POINT

> The universal quantifier is used when the associated predicate is *true* **for all possible** values of the quantified variable.

Worked examples

3.8 Express the following statement as a predicate: 'Every person managed by Jim is a programmer'; use the sets and functions from Example 3.3.

Solution $\forall e \in E : \text{Manages}(\text{Jim}, e) \Rightarrow \text{EmployedAs}(\text{programmer}, e)$

In this solution the universal quantifier is used to indicate that the implication that follows applies to **every** employee managed by Jim.

3.9 Express the fact that if every element of set S_1 is an element of set S_2 then S_1 is a subset of S_2. Use the sets and functions from Example 3.4.

Solution $(\forall e \in U : (\text{IsIn}(S_1, e) \Rightarrow \text{IsIn}(S_2, e))) \Leftrightarrow \text{SubsetOf}(S_2, S_1)$

In this solution the universal quantifier is used to indicate that the implication applies to all elements of the sets S_1 and S_2.

3.10 Given the following predicates defined for the set of integers $\mathbb{Z}$:
Odd: $\mathbb{Z} \rightarrow \mathbb{B}$ Indicates if a given integer is an odd number.
Even: $\mathbb{Z} \rightarrow \mathbb{B}$ Indicates if a given integer is an even number.
Prime: $\mathbb{Z} \rightarrow \mathbb{B}$ Indicates if a given integer is a prime number.
Divides: $\mathbb{Z} \times \mathbb{Z} \rightarrow \mathbb{B}$ Indicates whether or not the first integer is exactly divisible by the second.
Write the following statements as quantified predicates:

(a) All integers are either odd or even.
(b) The sum of any two odd integers is an even integer.

Solution (a) $\forall i \in \mathbb{Z}\text{: } \text{Odd}(i) \vee \text{Even}(i)$
(b) $\forall i, j \in \mathbb{Z}\text{: } \text{Odd}(i) \wedge \text{Odd}(j) \Rightarrow \text{Even}(i + j)$

The range of quantifiers

So far we have expressed quantified predicates in the following forms: either $\forall i \in I : P(i)$ or $\exists i \in I : P(i)$. In each of these cases the domain of the predicate has been the whole of the set I. It may be the case that we wish to restrict the domain of P to some subset of I. The above predicates could then be written as $\forall i \in I : R(i) : P(i)$ or $\exists i \in I : R(i) : P(i)$ where $R(i)$ is some specified range of values of i from the set I. For example we may write: $\forall i \in \mathbb{N} : i > 4\text{: } \sqrt{i} > 2$, which can be verbalized as: 'Every natural number greater than 4 has a square root greater than 2'. An example of an expression involving the existential quantifier is: $\exists i \in \mathbb{N} : 1 < i < 6 : i^2 > 20$, which may be verbalized as: 'There is a natural number between 1 and 6 whose square is greater than 20'. Here are more examples of the use of ranges in quantified predicates:

Worked example

3.11 Write the following statements as predicates:

(a) The square of every negative integer is positive.
(b) There is a natural number less than 10 which is both a prime and even.

Solution (a) $\forall i \in \mathbb{Z} : i < 0 : i^2 > 0$
(b) $\exists i \in \mathbb{N} : i < 10 : \text{Even}(i) \wedge \text{Prime}(i)$

Quantifiers and the ranges *true* and *false*

Let us consider the case in which the range of an existentially quantified predicate is empty. An example of this is the expression:

$$\exists x \in \mathbb{N}: x > 4 \wedge x < 2: P(x)$$

Since there is no natural number which is both less than 2 and greater than 4 the range in the above expression is certainly empty. The predicate $x>4 \wedge x<2$, which specifies the range, has a value *false*. The quantified predicate may therefore be written as:

$$\exists x \in \mathbb{N} : \mathit{false} : P(x)$$

We can make this expression general by letting x range over some arbitrary set X. We write:

$$\exists x \in X : \mathit{false} : P(x)$$

For this expression to have the value *true* there must be at least one value of the variable x from the specified range that makes the predicate $P(x)$ *true*. Since the specified range is empty there can be no such value of x and hence the expression can only have the value *false*.

KEY POINT

Any existentially quantified predicate over an **empty range** has the value *false*. We write:

$$\exists x \in X : \mathit{false} : P(x) \Leftrightarrow \mathit{false}$$

By convention we have the following less obvious rule for the universal quantifier:

KEY POINT

The value of a universally quantified predicate over an empty range is always *true*. We write:

$$\forall x \in X : \mathit{false} : P(x) \Leftrightarrow \mathit{true}$$

If the range of a quantified predicate is to be the whole of the predicate's domain then the expression that specifies the predicate's range must evaluate to *true* for all values from the predicate's domain. In this case we write:

$$\exists x \in X : \mathit{true} : P(x)$$

and

$$\forall x \in X : \mathit{true} : P(x)$$

We can abbreviate these expressions to:

$$\exists x \in X : P(x)$$

and

$$\forall x \in X : P(x)$$

Worked examples

3.12 Show that it is true that all natural numbers greater than 10 but less than 4 are prime.

Solution Statement: 'All natural numbers greater than 10 but less than 4 are prime'. Writing this statement as a predicate we get:

$$\forall i \in \mathbb{N} : 10 < i < 4 : \mathrm{Prime}(i)$$

Since the range $10<i<4$ is empty, by convention, the predicate is *true* and hence so is the above statement.

3.13 Show that there is no even natural number less than 0.

Solution Statement: 'There is an even natural number less than 0'. Writing this statement as a predicate we get:

$$\exists i \in \mathbb{N} : i < 0 : \mathrm{Even}(i)$$

Since there are no natural numbers less than zero the range is empty and so, by convention, this predicate is *false*, hence so is the statement.

Mixing quantifiers

It is not unusual to find that the representation of a particular statement requires a predicate that contains a mixture of existential and universal quantifiers. When creating such predicates it is important to ensure the quantifiers are written in the correct sequence. The subtle effect that the sequencing of quantifiers may have on meaning is demonstrated in the following example.

Worked examples

3.14 Using the sets and functions from Example 3.3 write the following statements as predicates.

(a) In every project team there is at least one systems analyst.
(b) Every project team has the same systems analyst.

Solution Notice importance of the order of the quantifiers in the following solutions:

(a) $\forall t \in PT : (\exists e \in E : \text{IsIn}(t, e) \wedge \text{EmployedAs(systems analyst}, e))$
(b) $\exists e \in E : (\forall t \in PT : \text{IsIn}(t, e) \wedge \text{EmployedAs(systems analyst}, e))$

Here are some more examples of statements containing both existential and universal quantifiers.

3.15 Using the functions from example 3.10 express the following statements as quantified predicates.

(a) For any given prime number there is an even number which is exactly divisible by the given prime.
(b) Every even number is divisible by a prime number.

Solution (a) $\forall x \in \mathbb{N}\text{: } (\text{Prime}(x) \Rightarrow \exists y \in \mathbb{N}\text{: } \text{Even}(y) \wedge \text{Divides}(y, x))$
(b) $\forall x \in \mathbb{N}\text{: } (\text{Even}(x) \Rightarrow \exists y \in \mathbb{N}\text{: } \text{Prime}(y) \wedge \text{Divides}(x, y))$

Self-Assessment Question 3.3

If no students turn up for an examination is it true to say that all students who took the examination passed?

Exercise 3.3

1. Express as predicates the following facts relating to the natural numbers.

   (a) There is a natural number whose square is greater than 16.
   (b) There is no natural number whose square is a prime number.

2. Using the sets and functions from Example 3.3 express the following statements as predicates:

   (a) There is a member of team2 who is a programmer with six years' experience.
   (b) There is no employee who is a member of both team1 and team2.

3. Construct statements in English that are equivalent to the following predicates.

   Note: Part (c) refers to the sets and functions from Example 3.4 while parts (d) and (e) refer to Example 3.3.

(a) $\exists x \in \mathbb{N} : x^2 = 2x$
(b) $\neg\exists x \in \mathbb{N} : x^2 > (x+1)^2$
(c) $\neg\exists e \in E :$ IsIn$(S_1, e) \wedge$ IsIn$(S_2, e) \Rightarrow \neg$Equal$(S_1, S_2)$
(d) $\exists e \in E :$ Manages(Jill, e) $\wedge$ Leader(team2, e)
(e) $\neg\exists e \in E :$ ExpertIn(COBOL, e) $\wedge$ ExpertIn(C, e)

4. Using the functions from Example 3.10 write the following as predicates:

   (a) Every integer that is divisible by 2 is even.
   (b) The square of any natural number is positive.
   (c) The sum of an odd and an even integer is odd.

5. With reference to the functions of Example 3.10 express the following predicates in English:

   (a) $\forall i \in \mathbb{Z}$: Odd$(i) \Rightarrow \neg$Divides$(i, 2)$
   (b) $\forall i, j \in \mathbb{Z}$: Odd$(i) \wedge$ Even$(j) \Rightarrow$ Even$(i * j)$
   (c) $\forall i \in \mathbb{Z}$:$\neg$(Odd$(i) \wedge$ Even(i^2))

6. With reference to Example 3.3 write the following statements as predicates.

   (a) Every employee who is an expert in C++ is a programmer.
   (b) The leader of any project team has at least three years' experience.

7. Write the following statements as predicates:

   (a) There is no prime number between 32 and 36.
   (b) Every natural number between 4 and 20 which is not a prime is exactly divisible by either 3 or 2.
   (c) There is a positive integer less than 10 which is both even and prime.

8. Write the following as predicates:

   (a) Every natural number that is not prime is divisible by a prime number.
   (b) There is no largest natural number.

9. Using the sets and functions from Example 3.3 write the following statements as predicates.

   (a) Every systems analyst is managed by a person who has at least five years' experience.
   (b) There is a project team, the members of which are all experts in OOD.

10. Express the following predicates in English.

   (a) $\forall x \in \mathbb{N}$: (Prime$(x) \Rightarrow \exists y \in \mathbb{N}$: Divides$(x, y) \wedge (y \neq 1 \vee y \neq x$
   (b) $\forall x \in \mathbb{N}$: (Even$(x) \Rightarrow \exists y \in \mathbb{N}$: Odd$(y) \wedge (y = x + 1)$)

3.4 Substitution

The process of substitution involves the replacement of a variable in a formula by an expression, a constant or some other variable. The replacement of the variable x by the term E is indicated by $[x\backslash E]$. If we wish to replace the variable x in the formula

$$x > 0 \wedge y > 0 \Rightarrow x + y < 2x + y$$

by the expression $(x + 1)$ we would write:

$$(x > 0 \wedge y > 0 \Rightarrow x + y < 2x + y)[x\backslash(x + 1)]$$

giving

$$(x + 1) > 0 \wedge y > 0 \Rightarrow (x + 1) + y < 2(x + 1) + y$$

Carrying out replacements in formulae that contain no quantifiers is quite straightforward. The presence of quantifiers however makes things somewhat trickier.

In formulas that contain quantifiers two classes of variable can be identified: those that are **bound** to quantifiers and those that are **free**. In the formula

$$\exists x : x = n^2 \wedge x + n = 2m$$

where x, n and m represent natural numbers the variable x is bound to the existential quantifier while both of the variables n and m are free. We may rename bound variables as long as the process of renaming does not change the bindings in the formula. For example if we were to rename the variable x as y the formula would become:

$$\exists y : y = n^2 \wedge x + n = 2m$$

which is equivalent to the original formula. However, renaming x as n we get:

$$\exists n : n = n^2 \wedge n + n = 2m$$

which does not have the same meaning as the original formula.

Here is another example of a consistent formula in which the variables range over the set $\mathbb{Z}$ of integers:

$$(\exists x : y = x + 1) \wedge y = 2x$$

The variable y is a free variable while the variable x appears to be both free and bound. There is no contradiction, however, since the two occurrences of x represent two distinct variables. In the left-hand term of the conjunction the x represents a variable which is bound by the existential quantifier, while in the second term the x represents a **different** free variable.

Substitutions which preserve the consistency of the formula may be performed on any of the free variables. Here is an example:

$$((\exists x : y = x + 1) \wedge y = 2x)[y\backslash n] \text{ (with } n \in \mathbb{Z})$$
$$\text{gives } (\exists x : n = x + 1) \wedge n = 2x$$

It is also possible to perform a substitution which destroys the consistency of the formula as is shown in the following example.

$$((\exists x : y = x + 1) \wedge y = 2x)[y\backslash x]$$
gives $(\exists x : x = x + 1) \wedge x = 2x$

This last example differs from the first in that it has resulted in a once free variable becoming bound. This situation may be avoided by first renaming the bound variable. To do this any letter may be used for this task as long as it is not being used to represent a free variable somewhere else in the formula.

Starting with the original formula

$$(\exists x : y = x + 1) \wedge y = 2x$$

rename the bound variable giving

$$(\exists z : y = z + 1) \wedge y = 2x$$

Perform the substitution

$$((\exists z : y = z + 1) \wedge y = 2x)[y\backslash x]$$

to get

$$(\exists z : x = z + 1) \wedge x = 2x$$

Here are some more examples of substitutions.

Worked example

3.16 Perform the following substitutions:

(a) $(y = 2n + 2 \wedge \exists n : n = y + 1)\ [y\backslash x^2]$
(b) $(y = 2n + 2 \wedge \exists n : n = y + 1)\ [n\backslash x]$
(c) $(y = 2n + 2 \wedge \exists n : n = y + 1)\ [y\backslash n]$

Solution (a) $x^2 = 2n + 2 \wedge \exists n : n = x^2 + 1$
(b) $y = 2x + 2 \wedge \exists n : n = y + 1$
(c) This substitution would result in a change in binding so first rename the bound variable to give:

$$(y = 2x + 2 \wedge \exists x : x = y + 1)\ [y\backslash n]$$

and now the substitution to give:

$$n = 2x + 2 \wedge \exists x : x = n + 1$$

It is possible to specify the simultaneous substitution of more than one variable, this process being known as multiple substitution. The simultaneous substitution of x by E and y by F is indicated by $[x,y\backslash E,F]$.

Multiple substitution may not have the same effect as repeated substitution.

Worked example

3.17 Perform the following substitutions.

(a) $(x + 2 = 2y)[x, y \backslash y, x]$

(b) $(x + 2 = 2y)[x \backslash y]\ [y \backslash x]$

Solution (a) $(x + 2 = 2y)[x, y \backslash y, x]$

Simultaneous substitution gives

$y + 2 = 2x$

(b) $(x + 2 = 2y)[x \backslash y]\ [y \backslash x]$

First substitution gives

$(y + 2 = 2y)[y \backslash x]$

Second substitution gives

$x + 2 = 2x$

Self-Assessment Question 3.4

Why is it that $(\exists x : x = x + 1) \wedge x = 2x$ is not consistent while $(\exists z : x = z + 1) \wedge x = 2x$ is consistent?

Exercise 3.4

1. Carry out the following substitutions:

   (a) $(\forall n :\ (n = 2i \Rightarrow \exists k :\ (k = 2j + 1 \wedge k = n + 1)))[i \backslash x]$

   (b) $(\forall n :\ (n = 2i \Rightarrow \exists k :\ (k = 2j + 1 \wedge k = n + 1)))[j \backslash k]$

   (c) $(\forall x :\ 0 < x < n^2 :\ (\exists i :\ i = x + 1 \wedge i \leq n^2))[n \backslash 2k + 1]$

   (d) $(\forall x :\ 0 < x < n^2 :\ (\exists i :\ i = x + 1 \wedge i \leq n^2))[n \backslash i]$

2. Perform the following substitutions.

   (a) Manages(Jim, Sarah)[Jim,Sarah\Sarah,Jim]

   (b) Manages(Jim, Sarah)[Jim\Sarah][Sarah\Jim]

   (c) $(x^2 + 2y + 2xy)[x, y \backslash a, b]$

   (d) $(x^2 + 2y + 2xy)[x \backslash a][a, y \backslash b, a]$

3.5 Rules of inference

In order to be able to reason about statements that are represented using the predicate calculus we will need a set of rules of inference as was the case with the propositional calculus. All the relations and rules of inference from the propositional calculus will apply to predicates when we do not have to consider the presence of quantifiers.

Basic rules of inference

We have already met two rules relating to the evaluation of quantified predicates over empty ranges. They are:

Rule 1 ∀ Over an empty range

$$\forall x : false : P(x) \Leftrightarrow true$$

Rule 2 ∃ Over an empty range

$$\exists x : false : P(x) \Leftrightarrow false$$

We shall now state some basic rules which form the definition of the existential and universal quantifiers. From these rules we will derive more complex rules of inference which can be used when reasoning about predicate calculus statements. For the existential quantifier we have:

Rule 3 ∃ Trading

$$\exists x \in X : R(x) : P(x) \Leftrightarrow \exists x \in X : R(x) \wedge P(x)$$

The expression $\exists x \in X : R(x) : P(x)$ states that there is a value for the variable x for which both the predicate $P(x)$ and the range predicate $R(x)$ are *true*. In other words both $R(x)$ **and** $P(x)$ are *true* for some x. We may represent this statement as $\exists x \in X : R(x) \wedge P(x)$. This leads to the equivalence expression in this rule.

Rule 4 ∃ Introduction

$$\frac{(R(x) \wedge P(x))[x \backslash a]}{\exists x \in X : R(x) : P(x)}$$

This rule states that if a predicate is true for a particular value from its domain, then the existential quantification of that predicate, over a range containing that particular value, is also true.

Rule 5 ∃ One-point (Leibnitz rule)

$$\frac{\exists x \in X : x = a : P(x)}{P(x)[x \backslash a]}$$

This rule states that if the value of a predicate existentially quantified over a range containing just one particular value, is true, then that predicate is true when applied to that particular value.

Rule 6 ∃ Remove

$$\frac{\exists x \in X : P(x)}{P(x)[x \backslash a]} \qquad \text{'}a\text{' fresh and } P(a) \Leftrightarrow true$$

This rule can be used to remove the existential quantifier by introducing an identifier for which the statement $P(x)$ is true. To avoid invalid inferences this new identifier must not have been used elsewhere.

The following rules relate to the universal quantifier:

Rule 7 ∀ Trading

$$\forall x \in X : R(x) : P(x) \Leftrightarrow \forall x \in X : R(x) \Rightarrow P(x)$$

The expression $\forall x \in X$: $R(x)$: $P(x)$ states that if a value of the variable x satisfies the range predicate $R(x)$ then the predicate $P(x)$ is *true*. This statement may be written as $\forall x \in X : R(x) \Rightarrow P(x)$. This leads to the equivalence expression in this rule.

Rule 8 ∀ Introduction

$$\frac{P(x)}{\forall x \in X : P(x)}$$

This rule states that if a predicate evaluated for any value from its domain is true, then the universal quantification of that predicate is also true.

Rule 9 ∀ Remove

$$\frac{\forall x \in X : P(x)}{P(x)[x \backslash a]}$$

This rule states that if a universally quantified predicate is true, then that predicate will be true for any element of its domain.

The final basic rule is one which connects the universal and existential quantifiers.

Rule 10 ∀∃ Relationship

$$\forall x \in X : P(x) \Leftrightarrow \neg(\exists x \in X : \neg P(x))$$

This rule states that if a universally quantified predicate is true then it is not the case that there is a value in its domain for which it is false, and vice versa.

Derived relations

We can now derive relations for the predicate calculus using the above rules. First we will consider the effect of negation on predicates. This will lead to a generalization of de Morgan's law.

Worked example

3.18 Justify the following relation which is the generalization of de Morgan's law for the universal quantifier.

∀ de Morgan $\qquad \neg(\forall x \in X : P(x)) \Leftrightarrow \exists x \in X : \neg P(x)$

Solution We need to show that the relation holds in both directions.

Firstly show that:

$$\frac{\neg(\forall x \in X : P(x))}{\exists x \in X : \neg P(x)}$$

Assertion	Reason
1. $\neg(\forall x \in X : P(x))$	Hypothesis
2. $\neg\neg(\exists x \in X : \neg P(x))$	1. $\forall\exists$ Relationship
3. $\exists x \in X : \neg P(x)$	2. Property of Negation

Secondly show that:

$$\frac{\exists x \in X : \neg P(x)}{\neg(\forall x \in X : P(x))}$$

Assertion	Reason
1. $\exists x \in X : \neg P(x)$	Hypothesis
2. $\neg\neg(\exists x \in X : \neg P(x))$	1. Property of Negation
3. $\neg(\forall x \in X : P(x))$	2. $\forall\exists$ Relationship

This concludes the justification.

3.19 Justify the following relation:

$\neg\forall$ Introduction

$$\frac{\neg P(x)[x\backslash a]}{\neg(\forall x \in X : P(x))}$$

Solution

Assertion	Reason
1. $\neg P(x)[x\backslash a]$	Hypothesis
2. $\exists x \in X : \neg P(x)$	1. $\exists$ Introduction
3. $\neg(\forall x \in X : P(x))$	2. $\forall$ de Morgan

This concludes the justification.

3.20 Justify the following relation:

$\neg\forall$ Removal

$$\frac{\begin{array}{c}\neg(\forall x \in X : P(x)) \\ \neg P(x)[x\backslash a] \Rightarrow Q\end{array}}{Q}$$

Solution

Assertion	Reason
1. $\neg(\forall x \in X : P(x))$	Hypothesis
2. $\neg P(x)[x\backslash a] \Rightarrow Q$	Hypothesis
3. $\neg(P(x)[x\backslash a])$	1. $\forall$ Remove
4. Q	3, 2. *modus ponens*

This concludes the justification.

Combining quantified predicates

Let us now consider the relationship between the quantifiers and the logical connectives $\wedge$ and $\vee$. For the universal quantifier we have the following relationships:

$\forall \wedge$ Distribution

$$(\forall x \in X : P(x)) \wedge (\forall x \in X : Q(x)) \Leftrightarrow \forall x \in X : (P(x) \wedge Q(x))$$

and

$\forall \vee$ Distribution

$$\frac{(\forall x \in X : P(x)) \vee (\forall x \in X : Q(x))}{\forall x \in X : (P(x) \vee Q(x))}$$

Worked example

3.21 Justify the relationship that describes the distribution of $\forall$ over $\vee$.

Solution

Assertion	Reason
1. $(\forall x \in X : P(x)) \vee (\forall x \in X : Q(x))$	Hypothesis
2. $\forall x \in X : P(x)$	Hypothesis
3. $P(a)$	2, $\forall$ Remove ('a' arbitrary)
4. $P(a) \vee Q(a)$	Addition ('a' arbitrary)
5. $\forall x \in X : (P(x) \vee Q(x))$	4, $\forall$ Introduction
6. $\forall x \in X : P(x) \Rightarrow$ $\forall x \in X : (P(x) \vee Q(x))$	2, 5
7. $\forall x \in X : Q(x)$	Hypothesis
8. Q(a)	7, $\forall$ Remove ('a' arbitrary)
9. $P(a) \vee Q(a)$	Addition ('a' arbitrary)
10. $\forall x \in X :(P(x) \vee Q(x))$	9, $\forall$ Introduction
11. $\forall x \in X : Q(x) \Rightarrow$ $\forall x \in X : (P(x) \vee Q(x))$	7, 10
12. $(\forall x \in X : P(x) \Rightarrow$ $\forall x \in X : (P(x) \vee Q(x))) \wedge$ $(\forall x \in X : Q(x) \Rightarrow$ $\forall x \in X : (P(x) \vee Q(x)))$	6, 11
13. $\forall x \in X : (P(x) \vee Q(x))$	1, 12, Constructive dilemma

This concludes the justification.

Self-Assessment Question 3.5

Can you formalize the statement that is represented by the phrase 'vice versa' in Rule 10?

Exercise 3.5

1. Justify the generalization of de Morgan's law for the existential quantifier.

   $\exists$ de Morgan

   $$\neg(\exists x \in X : P(x)) \Leftrightarrow \forall x \in X : \neg P(x)$$

2. Justify the following relations:

   (a) $\neg\exists$ introduction

   $$\frac{\neg P(x)}{\neg(\exists x \in X : P(x))}$$

   (b) $\neg\exists$ removal

   $$\frac{\neg(\exists x \in X : P(x))}{\neg P(x)[x\backslash a]} \qquad \text{'}a\text{' fresh and } P(a) \Leftrightarrow \mathit{true}$$

3. Justify the following relationship:

   $\forall\wedge$ distribution

   $$(\forall x : P(x)) \wedge (\forall x : Q(x)) \Leftrightarrow \forall x : P(x) \wedge Q(x)$$

The relationships between the existential quantifier and the connectives $\wedge$ and $\vee$ follow. The justifications of these relationships are left for later exercises.

$\exists\wedge$ Distribution

$$\frac{\exists x \in X : P(x) \wedge Q(x)}{(\exists x \in X : P(x)) \wedge (\exists x \in X : Q(x))}$$

$\exists\vee$ Distribution

$$\exists x \in X : (P(x) \vee Q(x)) \Leftrightarrow (\exists x \in X : P(x)) \vee (\exists x \in X : Q(x))$$

3.6 Reasoning with quantified predicates

We are now in a position to use the rules and relationships of the predicate calculus to analyse the validity of statements. The following examples and exercises relate to the sets and functions that first appeared in Example 3.3.

Worked examples

3.22 A company has the policy that any employee who is a team leader should not be a programmer and that every manager should be a team leader. Show that these policies imply that any employee who is a manager cannot be a programmer.

Solution The key phrases from the above description are:

Any team leader should not be a programmer
Every manager should be a team leader
Any manager is not a programmer

Translating the above statements into predicates we get the following hypotheses:

$$\forall e \in E : (\exists t \in PT : \text{Leader}(t, e) \Rightarrow \neg \text{EmployedAs}(\text{programmer}, e))$$

and

$$\forall e \in E : (\exists d \in E : \text{Manages}(d, e) \Rightarrow \exists t \in PT : \text{Leader}(t, e))$$

from which we wish to show that:

$$\forall e \in E : (\exists d \in E : \text{Manages}(d, e) \Rightarrow \neg \text{EmployedAs}(\text{programmer}, e))$$

Assertion	Reason
1. $\forall e \in E : (\exists t \in PT : \text{Leader}(t, e) \Rightarrow \neg \text{EmployedAs}(\text{programmer}, e))$	Hypothesis
2. $\forall e \in E : (\exists d \in E : \text{Manages}(d, e) \Rightarrow \exists t \in PT : \text{Leader}(t, e))$	Hypothesis
3. $\exists t \in PT : \text{Leader}(t, \text{Joe}) \Rightarrow \neg \text{EmployedAs}(\text{programmer}, \text{Joe})$	1, $\forall$ Remove
4. $\exists d \in E : \text{Manages}(d, \text{Joe}) \Rightarrow \exists t \in PT : \text{Leader}(t, \text{Joe})$	2, $\forall$ Remove
5. $\exists d \in E : \text{Manages}(d, \text{Joe}) \Rightarrow \neg \text{EmployedAs}(\text{programmer}, \text{Joe})$	4, 3, Hypothetical syllogism
6. $\forall e \in E : (\exists d \in E : \text{Manages}(d, e) \Rightarrow \neg \text{EmployedAs}(\text{programmer}, e))$	5, $\forall$ Introduction

This concludes the justification.

3.23 Here is another description of a company policy. Is it valid to draw the stated conclusion?

It is the company's policy that if any employee is a manager then he or she must be both a team leader and a systems analyst. Some of the managers in the company have more than ten years' experience. Consequently we may conclude that there are some team leaders who also have more than ten years' experience.

Solution The above description and associated conclusion may be written as:

Hypotheses:

$$\forall e \in E : (\exists d \in E : \text{ Manages}(d, e) \Rightarrow$$
$$\exists t \in PT : \text{Leader}(t, e) \wedge \text{EmployedAs}(\text{systems analyst}, e))$$
$$\exists e \in E : (\exists d \in E : \text{ Manages}(d, e) \wedge (\text{Experience}(e) > 10)$$

The conclusion that needs justifying is:

$$\exists e \in E : (\exists t \in PT : \text{ Leader}(t, e) \wedge (\text{Experience}(e) > 10))$$

Justifying the conclusion we have:

Assertion	Reason
1. $\forall e \in E : (\exists d \in E : \text{ Manages}(d, e) \Rightarrow \exists t \in PT : \text{ Leader}(t, e) \wedge \text{EmployedAs}(\text{systems analyst}, e))$	Hypothesis
2. $\exists e \in E : (\exists d \in E : \text{ Manages}(d, e) \wedge (\text{Experience}(e) > 10)$	Hypothesis
3. $\exists d \in E : \text{ Manages}(d, \text{Joe}) \wedge (\text{Experience}(\text{Joe}) > 10)$	2, $\exists$ remove 'Joe' fresh
4. $\exists d \in E : \text{ Manages}(d, \text{Joe})$	3, Simplification
5. $\text{Experience}(\text{Joe}) > 10$	3, Simplification
6. $\exists d \in E : \text{ Manages}(d, \text{Joe}) \Rightarrow (\exists t \in PT : \text{ Leader}(t, \text{Joe}) \wedge \text{EmployedAs}(\text{systems analyst}, \text{Joe}))$	1, $\forall$ Remove
7. $\exists t \in PT : \text{ Leader}(t, \text{Joe}) \wedge \text{EmployedAs}(\text{systems analyst}, \text{Joe})$	4, 6, *modus ponens*
8. $\exists t \in PT : \text{ Leader}(t, \text{Joe})$	7, Simplification
9. $\exists t \in PT : \text{ Leader}(t, \text{Joe}) \wedge (\text{Experience}(\text{Joe}) > 10)$	8, 5, Conjunction
10. $\exists e \in E : (\exists t \in PT : \text{ Leader}(t, e) \wedge (\text{Experience}(e) > 10))$	10, $\exists$ Introduction

This concludes the justification.

Exercise 3.6

1. Explain why the company description in Example 3.20 may also be represented by the following predicate statements and show that this alternative relation is also valid:

   Hypotheses:

   $$\neg \exists e \in E : (\exists t \in PT : \text{ Leader}(t, e) \wedge \text{EmployedAs}(\text{programmer}, e))$$

and

$$\forall e \in E : (\exists d \in E : \text{Manages}(d, e) \Rightarrow \exists t \in PT : \text{Leader}(t, e))$$

Conclusion:

$$\neg \exists e \in E : (\exists d \in E : \text{Manages}(d, e) \wedge \text{EmployedAs}(\text{programmer}, e))$$

2. In a particular company there is an expert in OOD who manages all of the systems analysts. The company has a policy that experts in OOD will not manage programmers. Can we conclude that no employee of the company who is an analyst is also a programmer?

Test and Assignment Exercises 3

1. Use the functions and sets from Example 3.4.
   (a) Write the following statements as predicates:
      (i) If every member of set S_1 is a member of set S_2 and if every member of set S_2 is a member of set S_1, then the two sets are equal.
      (ii) If two sets are equal then there is no element of one set that is not an element of the other.
   (b) Convert the following predicates into English:
      (i) $\text{IsIn}(S_1, e) \wedge \neg \text{IsIn}(S_2, e) \Rightarrow \neg \text{Equal}(S_1, S_2)$
      (ii) $\neg (\text{IsIn}(S_1 \cup S_2, e) \Rightarrow \text{IsIn}(S_1 \cap S_2, e))$
2. Write the following statements as predicates:
   (a) There is no number that is a factor of (i.e exactly divides) every square number.
   (b) Every square number has a factor other than itself or 1.
   (c) There is only one even prime number.
3. Convert the following predicates into English
   (a) $\forall x \in \mathbb{N}\text{: Odd}(x) \vee \text{Even}(x)$
   (b) $\forall x \in \mathbb{N}\text{: } (\neg \text{Prime}(x) \Rightarrow \exists y \in \mathbb{N} : y \neq 1 \wedge y \neq x \wedge \text{Divides}(x, y))$
   (c) $\neg \exists x \in \mathbb{N}\text{: } 37 \leq x \leq 48 : x = n^2$
4. Using the sets and functions from Exercise 3.3 write the following statements as predicates.
   (a) Every manager is a systems analyst.
   (b) Every systems analyst is an expert in OOD.
   (c) Every employee who is an expert in OOD is a systems analyst.
   (d) There is a member of staff with more than ten years' experience who is a manager of all team leaders.

   In order to answer the following questions you will need to invent appropriate sets and predicates.
5. All employees who are experts in formal methods can produce consistent specifications for any given problem that requires specification. There are problems that need specifying. Any employee who produces a consistent specification for a problem receives a bonus. Jean is an expert in formal methods. Show that it is valid to assume that Jean will receive a bonus.
6. A consultant is employed to check on the practices of a software company. She is told that the company works on the premise that any module that satisfies a specification will pass all test procedures. The company has a policy that any module that passes all test procedures will be released. When the consultant is told that some modules have already been released she expresses the view that there could exist a module for which it is wrong to conclude that because it has been released it satisfies the specification. Use the predicate calculus to produce an argument to support the view held by the consultant.

3.7 Further reading

D C Ince, *An Introduction to Discrete Mathematics and Formal System Specification* (Clarendon, 1992)

An introductory text covering both the propositional and predicate calculus before going on to deal with the specification of simple systems. This text contains many simple concrete examples.

Jim Woodcock and Martin Loomes, *Software Engineering Mathematics* (Pitman, 1988)

Another introductory text which starts by introducing the propositional and predicate calculus before going on to look at the use of Z notation for specifying software systems.

Cliff B Jones, *Systematic Software Development Using VDM* (Prentice-Hall International, 1989)

A more advanced text than the previous two, covering predicate calculus proofs and the use of VDM in specifying software systems.

4 Mathematical proofs

Objectives

When you have mastered this chapter you should be able to

- read and construct proofs using each of the techniques discussed
- read and construct proofs using combinations of the techniques discussed

We now consider explicitly some of the common techniques used to prove statements. We examine some standard forms of proof, and in particular consider proof by induction, which allows us to prove infinitely many statements with a finite amount of work. Most of the actual techniques to be discussed here – with the exception of induction – have already been seen in the chapters on propositional and predicate calculus. However, we will now be seeing how these constructions are used to write proofs in a rather less formal way than before. This will provide familiarity with one of the ways in which mathematical thinking is used in practice, to provide reliable arguments without going all the way along the path of constructing formal proofs.

4.1 Direct proof

A very common type of proof is a chain of statements, each of which implies the next, where the first is something we know to be true, and the last is what we want to prove. In symbolic logic terms, we have A, a proposition which is known to be true, and a chain of implications of the form $A \Rightarrow B$, $B \Rightarrow C$ and so on, up to $Y \Rightarrow Z$. Repeated application of the *modus ponens* rule shows us that Z is a consequence of this. A proof of this sort is sometimes called a **chain of inference**.

Worked example

4.1 Prove that if S is the set of one and two digit numbers in which each of the digits 0 to 9 occurs exactly once, then the sum of the numbers in S is a multiple of 9.

Solution Since each of the numbers has either one or two digits, each of 0 to 9 must occur in either the units or the tens place. Denote by N the sum of the numbers in the units place. The numbers from 0 to 9 add up to 45, and so the sum of the numbers in the tens place must be $45 - N$. From this, we see that the sum of the numbers in S is $10(45 - N) + N$, which is equal to $450 - 9N$. 450 is a multiple of 9, and so is $-9N$. Thus $450 - 9N$ is also a multiple of 9, *i.e.* the sum of the numbers in S is a multiple of 9.

Note This can be applied to the problem of proving that a program satisfies a specification as follows. Suppose that P is a proposition describing the data to be fed to a program M, that Q is a proposition describing the desired output, and that the program consists of modules $M_1, \ldots, M_n$. Then the output from each module is the input for the next. Now, use $A \overset{M_i}{\Rightarrow} B$ to mean that whenever the input for M_i satisfies the condition A, the output will satisfy the condition B. If we can find conditions $A_1, \ldots, A_{n-1}$ such that $P \overset{M_1}{\Rightarrow} A_1$, $A_1 \overset{M_2}{\Rightarrow} A_2, \ldots,$ $A_{n-1} \overset{M_n}{\Rightarrow} Q$, then we know that $P \overset{M}{\Rightarrow} Q$, which is what we wanted to prove.

The point of this is that rather than having to solve the big problem, $P \overset{M}{\Rightarrow} Q$, all at once, we can reduce it to a collection of simpler problems, solve them, then reconstruct the proof we actually want.

Self-Assessment Question 4.1

What is the relation between this kind of proof and the formal proofs that we met in the chapters on propositional and predicate calculus?

Exercise 4.1

Charles is joining a two-man team, currently consisting of Albert and Bertram. He has been warned that one of them is a compulsive liar, and discovers to his horror that neither of them is wearing an identity card. One of them, however, is wearing a badge that says 'team manager'. He asks the pair 'Is Albert a compulsive liar?', and the one with the 'team manager' badge answers either 'yes' or 'no'. From this answer, he is able to work out which of them is Albert. What was the answer, and which one is Albert?
Hint: The team manager may have said either 'yes' or 'no' and be a compulsive liar, or not. Consider each possibility in turn, and see which of them allows you to make a deduction.

4.2 Counter-example

Often, one wishes to check the validity of a statement of the form $\forall x : P(x)$. Before attempting to prove such a statement, one should briefly check that there are no obvious **counter-examples**, *i.e.* no values of x for which $P(x)$ is false. Since we know that $\neg \forall x : P(x)$ is logically equivalent to $\exists x : \neg P(x)$, we see that the existence of a single example for which the statement P fails to be true is enough to tell us that the universally quantified statement must be false.

Worked example

4.2 Is the statement 'All prime numbers are odd' true?

Solution Since 2 is a prime number, and is even, it is not the case that all prime numbers are odd.

Note that failure to find a counter-example is not enough to prove the statement true! It may just be that the attempt to find the counter-example was insufficiently ingenious. We should also note that many particular cases where the statement is true does not prove that it is always true. Only one counter-example is needed to show that a universally quantified statement is false, whereas to prove it true, a proof is required that covers all cases.

In fact, there are many statements in mathematics where nobody has yet succeeded in finding a counter-example, or a proof of the general case. Famous examples are provided by number theory: Goldbach's conjecture states that every even number can be expressed as the sum of two primes; all known perfect numbers are even, though there is no proof that there are no odd ones (a number is perfect if it is equal to the sum of its proper divisors). There is also the famous Fermat 'theorem', which says that there are no solutions of $a^n + b^n = c^n$ for non-zero integers a, b, c and n, where $n > 2$.

Self-Assessment Question 4.2

I claim that whenever you square an even number, you get a multiple of 4; as evidence for my claim, I present you with a table of the squares of all even numbers from 2 to 100, and show that each of them is a multiple of 4. Why is what I have done not a proof?

Exercise 4.2

Show by means of a counter-example that the statement 'The sum of two prime numbers is never a prime number' is false.

4.3 Contradiction and the contrapositive

Occasionally it is hard to prove the desired statement, but surprisingly easy to prove a differently phrased but equivalent one. One particular example of this is when we are required to show that $P \Rightarrow Q$. (In the context of program proof, we may wish to show that if P is true before the program runs, then Q is true afterwards.) Sometimes, when it is difficult to see how to establish the given proposition, replacing it by the equivalent **contrapositive** form $\neg Q \Rightarrow \neg P$ can make life easier.

Worked example

4.3 Show that if three numbers a, b and c are such that $a + b + c > 3$, then at least one of a, b or c must be greater than 1.

Solution After a few attempts to prove this directly, I come to the conclusion that I cannot do it, and so attempt to use the contrapositive form of the statement. This says that if all of a, b and c are no greater than 1, then $a + b + c \leq 3$. But this is obvious, since if $a \leq 1$, $b \leq 1$ and $c \leq 1$, then $a + b + c \leq 1 + 1 + 1 = 3$. Since we have proved this, we can deduce that if $a + b + c > 3$, at least one of a, b or c is greater than 1.

A special case of this is when we want to prove P, but cannot. We observe that $P \Leftrightarrow (T \Rightarrow P)$, and note that the contrapositive of this is $\neg P \Rightarrow F$. So if we want to prove that P is true, we can do this by assuming that P is false, and showing that this implies something known to be false. This strategy is called **proof by contradiction** and is one of the most powerful tools in the mathematician's kit.

Worked example

4.4 Show that there are no integers, m and n, such that $\sqrt{2} = m/n$.

Solution Let us suppose that m and n are indeed integers such that $\sqrt{2} = m/n$. We can assume that any common factors have been cancelled out. Now, if we square m/n we obtain 2, so that $m^2 = 2n^2$, and so m^2 must be even. It follows that m must also be even, for if m were odd, m^2 would be odd, therefore m must be even, *i.e.* $m = 2m'$, for some m'. But now we have $2n^2 = 4m'^2$, so that $n^2 = 2m'^2$ and n^2 is also even. But just as before, this tells us that n must be even. But now we have the consequence that two numbers with no common factors are both even, and hence have a common factor of 2. This provides the required contradiction, and demonstrates that $\sqrt{2}$ cannot be expressed as the ratio of two integers with no common factors.

Self-Assessment Question 4.3

Suppose that you try to prove some statement, P, by contradiction. You assume $\neg P$, but fail to deduce from this assumption any statement known to be false. Does this prove that P is not in fact true?

Exercise 4.3

1. I have a program which I wish to test for correctness. I know that if it is correct, then when I run it with an input value of 4, the output should be 10. To test it, I run it with input value 4 and it does indeed return an output of 10. Is there anything wrong with my argument when I say that, using the contrapositive, this shows that the program is correct?
2. A forgetful programmer has three different versions of a program, one on each of the computers named Tom, Jerry and Butch. He wants to use the most recent version, but cannot remember which one it is. He can remember that it is either the Tom version or the Jerry one, and that either the Tom version is oldest or the Butch version is most recent. Which version is most recent?

   Hint: Use proof by contradiction to eliminate two of the possibilities.

4.4 Induction

Sometimes we have a collection of statements, one about each integer, and we want to prove that they are all true. Since no matter how many of the statements we prove there will still be more left, we cannot prove them all simply by considering each case separately; a better method must be found. One technique used in such circumstances is **mathematical induction**. The basic idea is a simple one, though it often seems a little strange to begin with.

So suppose we have a collection of propositions, $P(1), P(2), P(3), \ldots$, and we wish to prove that they are all true. Clearly, we cannot prove them one by one, because we will never finish; a better method must be found. However, suppose we can show that for any value of k, $P(k) \Rightarrow P(k+1)$, and furthermore, we can prove that $P(1)$ is true.

Then we know that $P(1)$ is true, and we also know that $P(1) \Rightarrow P(2)$. That means that we know that $P(2)$ is true. But we also know that $P(2) \Rightarrow P(3)$, so then $P(3)$ is true, and so on. Whatever number you think of, eventually we get to it by this process, so we have a way of proving that $P(n)$ is true for any value of $n \in \mathbb{N}$ whatever.

A proof that proceeds by the above method is called a **proof by induction**. The proof that $P(1)$ is true is called the **basis for induction**, and the proof that $P(k) \Rightarrow P(k+1)$ for all k is called the **inductive step**.

Worked example

4.5 A sorting algorithm requires one operation to sort a list containing a single item. If the number of items in the list is k, then adding another item increases the number of operations by $2k + 1$. Show that n^2 operations are required for this algorithm to sort a list of n items.

Solution Let $P(n)$ be the statement 'it takes n^2 operations to sort n items'. We must establish the basis for induction, then prove the inductive step.

First, $P(1)$ says that it takes 1 operation to sort a list of 1 item. This is true, so we have the inductive basis.

Next, we wish to show that if $P(k)$ is true, then so is $P(k+1)$. So let us assume $P(k)$ for the moment. From this, the number of steps to sort k items is k^2. Then the number of steps required to sort $k+1$ items is $k^2 + 2k + 1$, which is simply $(k+1)^2$. Thus $P(k+1)$ is implied by $P(k)$. This gives us the inductive step.

We can conclude that $P(n)$ is true for all values of n.

We can often use mathematical induction to prove that certain sums are given by formulae.

Worked example

4.6 Prove that

$$\frac{1}{1 \times 2} + \frac{1}{2 \times 3} + \frac{1}{3 \times 4} + \cdots + \frac{1}{n \times (n+1)} = 1 - \frac{1}{n+1}$$

Solution Use $S(n)$ to represent the sum shown above. Then first, we must check that the formula gives $S(1)$ correctly. Putting $n = 1$ we obtain $1/2$ for the sum, and from the formula. Next, suppose that the formula gives $S(k)$ correctly for some arbitrary integer k. Then

$$\begin{aligned}
S(k+1) &= S(k) + \frac{1}{(k+1) \times (k+2)} \\
&= 1 - \frac{1}{k+1} + \frac{1}{(k+1) \times (k+2)} \\
&= 1 - \frac{k+2-1}{(k+1) \times (k+2)} \\
&= 1 - \frac{k+1}{(k+1) \times (k+2)} = 1 - \frac{1}{k+2}
\end{aligned}$$

which is the formula we require. Hence, by induction, the formula works for any positive integer n.

In fact, sums of this sort are sufficiently common that there is a special notation for them. We use the notation

$$\sum_{r=1}^{n} F(r) = F(1) + F(2) + \cdots + F(n)$$

It is well worth becoming acquainted with this notation. The first thing to observe is that there is nothing special about the use of the letter r in the above; when the sum is written out in full, all that appears is the range of values that r takes on.

Worked example

4.7 Write out each of the following sums in full.

$$\sum_{k=1}^{5} k^2 \qquad \sum_{r=1}^{4} r^2 \qquad \sum_{n=1}^{7} \frac{1}{n}$$

Solution

$$\sum_{k=1}^{5} k^2 = 1 + 4 + 9 + 16 + 25 \qquad \sum_{r=1}^{4} r^2 = 1 + 4 + 9 + 16$$

$$\sum_{n=1}^{7} \frac{1}{n} = 1 + \frac{1}{2} + \frac{1}{3} + \frac{1}{4} + \frac{1}{5} + \frac{1}{6} + \frac{1}{7}$$

Application to programming

We can use induction as a tool in the attempt to show that a loop does what it is supposed to do. For the purposes of this section, we consider only `while` loops, which we will represent as

```
while <cond> do
  begin
  S
  end
```

where `<cond>` is some Boolean expression and `S` is a section of code.

Now, suppose that `E` is some expression in the variables of the program in which the loop lies. If `E` has the same value before the loop executes as it does after the body of the loop has been executed once, then it will have the same value however often the body of the loop is executed. Such an expression is sometimes called a **loop invariant**. This fact can be used to analyse the behaviour of a loop as we now see.

Worked example

4.8 Consider the program segment

```
x:=m; y:=n;
while y > 0 do
  begin
  x:=x+1;
  y:=y-1;
  end
```

Show that the value $x + y$ is unchanged when the loop body is executed, and deduce that when the loop terminates the value stored in x is $m + n$, provided that m and n are positive integers.

Solution We use x and y to denote the values of x and y before the body of the loop is executed, and x' and y' to denote the values after. Then $x' + y' = x + 1 + y - 1 = x + y$, and so the value of $x + y$ is unchanged when the body of the loop is executed once. It follows that it is unchanged however often the body of the loop is executed.

Now, initially, $x + y$ has the value $m + n$, so this is also the value when the loop terminates. Furthermore, when the loop terminates $y = 0$, and so $x + 0 = m + n$, i.e. $x = m + n$.

Self-Assessment Question 4.4

Some students at first find proof by induction confusing, and argue as follows.

> Proof by induction doesn't really prove anything, because in the inductive step, one makes the assumption that $P(k)$ is true. Since one has to do this for each k, one has assumed what was to be proved in the first place, and so the proof method is invalid.

What is wrong with that argument?

Exercise 4.4

1. In a certain software house, each programming team leader must have a direct phone line to each of the others. Use induction to prove that if there are n team leaders, then $\frac{1}{2}n(n-1)$ lines are required.
2. Write each of the following sums out in full.

$$\sum_{m=1}^{6} k^{-2} \qquad \sum_{l=1}^{6} \frac{1}{l^2} \qquad \sum_{n=1}^{4} \frac{n}{n+1}$$

3. Use induction to show that if E is unchanged by one execution, it will be unchanged by any number of executions.
4. Show that $x.2^y$ is unchanged by the execution of the body of the loop in the following segment of code, where n is a positive integer.

```
x:=1; y:=n;
while y>0 do
  begin
  x:=2*x;
  y:=y-1;
  end
```

Deduce that when the loop terminates, the value of x is 2^n.

Test and Assignment Exercises 4

1. There is a mythical island all of whose inhabitants are either knights (who always tell the truth) or knaves (who always lie). Analysing the statements made by the inhabitants gives useful practice in using the logical tools used to analyse and prove statements.

(a) On your annual vacation on the island of knights and knaves – a location about which nobody else knows –

you meet Alice and Bartholomew. Alice says 'If I am a knight, then Bartholomew is a knave'. Bartholomew says 'Alice and I are both knights'. Is Alice a knight or a knave? What about Bartholomew?

(b) The next day it is raining, and you meet Albert and Barbara. Albert says 'I am a knight. It is raining today', and Barbara says 'I am a knight and it is raining today'. What can you say about Albert and Barbara?

(c) Later that day, you overhear Anne saying 'I'd tell the truth if I were you' to Benjamin. What can you deduce about Anne and Benjamin?

(d) In the evening, you make the mistake of going to a rather sleazy nightclub where you meet Zebedee, who announces with a leer, 'I am a knave, and women find me irresistible'. What conclusions can you draw about Zebedee?

(e) The next year, the island of knights and knaves has been discovered as a holiday spot. This horrifies you, but since you can now take a package tour at a fraction of the previous cost, you decide to visit the island again anyway. You meet Xavier, who tells you 'I am a knave'. What can you deduce about Xavier?

2. Find a counter-example for each of the following statements:

(a) All prime numbers are odd.
(b) All odd numbers are prime.

3. Use induction to prove each of the following statements.

(a) $1 + 2 + \cdots + n = \frac{1}{2}n(n+1)$ if $n \geq 1$
(b) $1^2 + 2^2 + \cdots + n^2 = \frac{1}{6}n(n+1)(2n+1)$ if $n \geq 1$
(c) $1.2 + 2.3 + \cdots + n(n+1) = \frac{n}{3}(n+1)(n+2)$ if $n \geq 1$
(d) $2^n > n^2$ if $n \geq 4$
(e) $n! > 2^n$ if $n \geq 4$

4. Consider the following section of code:

```
x:=0; y:=n;
while y>0 do
  begin
  x:=x+m;
  y:=y-1;
  end
```

Show that the value of $x + my$ is unchanged when the loop is executed, and deduce that when the loop halts x contains the value mn.

5. Consider the following section of code:

```
x:=m; y:=n;
while y>0 do
  begin
  x:=x-1;
  y:=y-1;
  end
```

Show that the value of $x - y$ is unchanged when the loop is executed, and deduce that when the loop halts x contains the value $m - n$.

4.5 Further reading

A Cupillari *The Nuts and Bolts of Proofs* (Wadsworth, 1989)
A presentation of techniques used to construct proofs, concentrating on traditional mathematical applications.
A Kaldewaij, *Programming: The Derivation of Programs* (Prentice Hall International, 1990)
E P Northrop *Riddles in Mathematics* (Penguin, 1960)
A fascinating collection of paradoxes and fallacious proofs.

5 Finite state automata

Objectives

When you have mastered this chapter you should be able to

- state what is meant by a language, and in particular a regular language
- construct automata to carry out simple tasks
- construct a deterministic finite state automaton (FSA) to do the same job as a given non-deterministic one
- find the regular expression describing the language accepted by a given FSA
- find an FSA accepting the language defined by a regular expression

The finite state automaton (often abbreviated to FSA) is one model of what is meant by a computer. In this chapter we investigate some variations on this idea and observe that they are all equivalent. We also consider automata as machines for recognizing languages and note that automata recognize precisely the languages described by regular expressions.

5.1 Languages

In general, a finite state automaton is a simple kind of machine, which may be used for various different tasks. We will consider it primarily as a machine for recognizing certain combinations of symbols – *i.e.* as a pattern matching device. We will then briefly consider some extensions of the idea. But it is convenient to have a brief digression on the mathematical notion of a language, in order to have a way of talking about these combinations of symbols and a concise way of describing certain types of language.

So, mathematically speaking, an **alphabet** is a non-empty set, Σ, whose elements are called **symbols**. A **string** over the alphabet Σ consists of a finite ordered collection of elements of Σ, which may have repetitions; the **empty string** is the string with no symbols at all in it, and is denoted ε. The set of all strings over Σ is denoted Σ^*; this $*$ operator is sometimes called the **Kleene star**. A **language** over Σ is any non-empty set of strings over Σ, and so is a subset of Σ^*.

Example

5.1 If $\Sigma = \{a, b, c\}$ then *aabc*, *bacab* and *aaaa* are all strings over Σ; *a*1*b* and *bad* are not, since they contain symbols that are not in the alphabet. This collection of three strings over Σ define a language (though a rather small one). Other possible languages are

- All strings consisting of just the letter *a*, repeated any number of times.
- All strings containing the same number of *a*s, *b*s and *c*s.
- All strings that are **palindromes** (*i.e.* they look the same whether you write them forward or backward, such as *abcba* or *aabccbaa*).

There are some useful operations and concepts associated with strings. For example, if Z is a string over Σ, then $\#(Z)$ is the **length** of Z and is the number of symbols occurring in Z (counted with repetitions).

Example

5.2 If Σ consists of $\{a, b, c, d\}$, then $\#(aabcad) = 6$, $\#(aaa) = 3$, and $\#(\varepsilon) = 0$.

If we have two strings, Z_1 and Z_2, then the **concatenation** of Z_1 and Z_2 is denoted Z_1Z_2, and is the string obtained by writing out the string Z_1 followed by Z_2. To give a slightly more concise way of writing, we use index notation in the following way: $Z^2 = ZZ$, $a^2b^3 = aabbb$, $(ab)^2 = abab$ and so on.

Example

5.3 With Σ as in Example 5.2, the concatenation of *aabc* and *bdc* is *aabcbdc*. Note that the concatenation of *bdc* with *aabc* is *bdcaabc*, so that concatenation is *not* a commutative operation.

We say that A is a **substring** of Z if the string A occurs inside Z (or is equal to it). More precisely, A is a substring of Z if there exist (possibly empty) strings C and B such that $Z = CAB$.

Example

5.4 If $A = aab$, then A is a substring of *abaaba*, *aababa* and *aaab*, but not of *ababab* or *abaa*.

There are several obvious but useful properties of strings and lengths, for example,

1. $\#(Z) \geq 0$ for any string Z over any language Σ.
2. $\#(Z) = 0$ if and only if $Z = \varepsilon$.
3. $\#(Z_1Z_2) = \#(Z_1) + \#(Z_2)$.
4. $Z\varepsilon = \varepsilon Z = Z$ for any Z.

Self-Assessment Question 5.1

Explain why if $AZ = Z$, then A must be ε.

Exercise 5.1

1. If $A = abc$ and $B = aabbb$, then write out A^2, ABA. Check that $\#(A^2) = 2\#(A)$ and that $\#(ABA) = 2\#(A) + \#(B)$.
2. Write out $a^3b^2a^2$, $(ab)^2$ and a^2b^2 in full.
3. Write *aaabbaaabbaaaabb* and *ababababab* using indices to compress the formulae.

5.2 Regular languages

There is a useful way of describing certain languages, known as the **regular languages**. They are described in terms of the pattern that strings must conform to in order to belong, and the same pattern matching ideas crop up in some editing and searching techniques.

A **regular expression** over Σ is defined as follows:

KEY POINT

1. ε is a regular expression.
2. If $x \in \Sigma$, then x is a regular expression.
3. If α and β are regular expressions, so are $\alpha\beta$ and $(\alpha \vee \beta)$.
4. If α is a regular expression, so is $(\alpha)^*$.
5. Nothing else is a regular expression.

In addition to these rules, we adopt one more convention to simplify the form of a regular expression; we replace $((\alpha))^*$ by $(\alpha)^*$ if that is possible, and if $x \in \Sigma$ we write x^* instead of $(x)^*$.

Note: This simply tells us what a regular expression looks like; we will see what they mean below. However, the * in this is related to the * in Σ^*, and the $\vee$ is related to the $\vee$ in symbolic logic because both are related to $\cup$ in set theory. We will see this more clearly later.

Worked example

5.5 Show that each of the following are regular expressions over $\{0, 1\}$.

1. $(1 \vee 0)^*01(01)^*$
2. $0^*(0 \vee 1^*)^*$

Solution

1. Since 0 and 1 are regular expressions, so are 01 and $(1 \vee 0)$. Since they are regular expressions, so are $(01)^*$ and $(1 \vee 0)^*$, and so is $(1 \vee 0)^*01(01)^*$.
2. Since 1 is a regular expression, so is 1^*. Since 0 is a regular expression, so are $(0 \vee (1)^*)$ and 0^*. Since $(0 \vee 1^*)$ is a regular expression, so is $(0 \vee 1^*)^*$, and thus so is $0^*(0 \vee 1^*)^*$.

Now, what language is defined by a given regular expression? We build the language in just the same way as we build the expressions, so that the following is true:

KEY POINT

1. ε corresponds to $\{\varepsilon\}$.
2. x corresponds to $\{x\}$.
3. If α corresponds to the language L_α and β to the language L_β, then $\alpha\beta$ corresponds to $\{st | s \in L_\alpha \wedge t \in L_\beta\}$ and $(\alpha \vee \beta)$ to $L_\alpha \cup L_\beta$.
4. If α corresponds to L_α, then $(\alpha)^*$ corresponds to L_α^*.

Worked example

5.6 Describe the language defined by each of the regular expressions

1. $(0^*10^*10^*)^*$
2. $(010 \vee 100 \vee 001)^*$

In the second case, re-introduce brackets to make the expression conform to the strict definition of a regular expression.

Solution

1. Any string with an even number of 1's.

2. Any string composed of the concatenation of strings of length 3, each of which has exactly one 1 in it. One regular expression which would do the job is $((010 \vee 100) \vee 001)^*$.

Self-Assessment Questions 5.2

What regular expression over the usual alphabet would describe any word ending in 'ize'? How could ideas like this be used in editing?

Exercise 5.2

1. Show that each of the following is a regular expression over the alphabet $\{a, b\}$.
   (a) $(ab \vee aba)^*$
   (b) $(a \vee (ab)^*)^*$
   (c) $((ab)^* \vee (ba)^*)$
2. Find regular expressions for each of the following languages.
   (a) Strings over $\{a, b, c\}$ of the form any number of a's, followed by any even number of b's, followed by either any odd number of c's or any number of a's.
   (b) Strings over $\{0, 1\}$ representing non-zero binary numbers that are divisible by 8.

5.3 Finite state automata

Now that we have seen a little of the theory of languages, we will proceed to look at one model of a machine that carries out computations. The model we will consider is a fairly simple one, known as a **finite state automaton**, and is only capable of a rather restricted type of computation.

Definition of a finite state automaton

We will begin by considering an everyday object which provides some motivation for the notion of an FSA. This object is the vending machine.

Think of what happens as you put money into a vending machine; as you put coins in, the machine keeps track of how much money you have inserted (it changes its internal state), until you have inserted a critical amount, allowing you to choose a product. When you make your choice, it drops out the item you have chosen, possibly with some change, and reverts back to the state it was in when you arrived. So the key notions here are of the internal state of the machine, and the inputs (in this case, either money, or a choice of product). From this we can abstract a mathematical model of an automaton. However, we will initially refrain from considering the possibility that the machine may give some output as well, and will restrict ourselves to that of the machine being in a satisfied state (as when enough money has been inserted to entitle the customer to choose an item).

KEY POINT

An FSA consists of a collection of 5 objects, namely a set S of states, a set Σ of **input symbols**, a function $d : S \times \Sigma \to S$, called the **transition function** (and which determines what happens to the state of the machine when it receives a given input), a special state s_0 called the **initial state**, and a subset A of S called the set of **accepting states**. You will sometimes see an FSA defined as a 5-tuple (meaning a collection of 5 objects in a particular order) (S, Σ, d, s_0, A).

Given all this information, we can see what happens to an FSA when a given string is presented to it; it starts in the initial state, then receives a symbol, say σ_1. The transition function tells us that the FSA then moves to state $d(s_0, \sigma_1)$. On processing the second symbol in the string, σ_2, the machine goes to state $d(d(s_0, \sigma_1), \sigma_2)$, and so on. We say that the machine **accepts** a string if it ends up in an accepting state, and that the machine **recognizes** or **accepts** a language if it accepts all the strings in the language, and no others.

Example

5.7 Consider the machine with two states, s_0 and s_1, where $A = \{s_1\}$, $\Sigma = \{a, b\}$, and the transition function d is given by $d(s_0, a) = s_0$, $d(s_0, b) = s_1$, $d(s_1, a) = s_1$, and $d(s_1, b) = s_0$. Then the input a makes no difference to the state of the machine, and the input b changes it from whichever state it is in to the other. Since the machine is in an accepting state when it is in state s_1, the machine accepts precisely those strings with an odd number of b's.

It is useful to have a way of following the behaviour of an FSA for a given input string; one way is to write the string out repeatedly, labelling each copy by the current symbol and state as the FSA works through it.

Worked example

5.8 Trace the behaviour of the FSA from Example 5.7 on the two strings *abbaaba* and *bbabb*.

Solution 1.

$$\underline{a}bbaaba : s_0$$
$$a\underline{b}baaba : s_0$$
$$ab\underline{b}aaba : s_1$$
$$abb\underline{a}aba : s_0$$
$$abba\underline{a}ba : s_0$$
$$abbaa\underline{b}a : s_0$$
$$abbaab\underline{a} : s_1$$
$$abbaaba : s_1$$

The last line shows the situation when the FSA has processed all the symbols in the string; since it is in state s_1, the string is accepted.

2.

$$\underline{b}babb : s_0$$
$$b\underline{b}abb : s_1$$
$$bb\underline{a}bb : s_0$$
$$bba\underline{b}b : s_0$$
$$bbab\underline{b} : s_1$$
$$bbabb : s_0$$

In this case, the FSA stops in state s_0, and so the string is not accepted.

Describing a finite state automaton

The formal definition of an FSA tells us what we need to say to specify one; it does not, however, insist that we say it in just that way. There are a couple of other ways of describing an FSA which are easier to work with, which we will now examine.

Pictorial descriptions of FSA

One possibility is just to draw a diagram of what the FSA does; this will have a circle to represent each state. To represent the transitions, if a given input causes the machine to go from one state to another we draw a line from the first state to the second, labelled by the input causing the transition, and with the arrow showing the direction of transition. (Such a representation is called a **directed graph**.) We point out the initial state by having an arrow pointing at it, and the accepting states by drawing a smaller circle inside the main one. This representation of an FSA is particularly useful for people, and it is easy to trace an FSA's behaviour on a given input simply by tracing through the diagram. It also seems to be the easiest one to use when constructing an FSA to carry out a particular task.

Example

5.9 The diagram for the automaton that recognizes strings with an odd number of b's is shown in Figure 5.1.

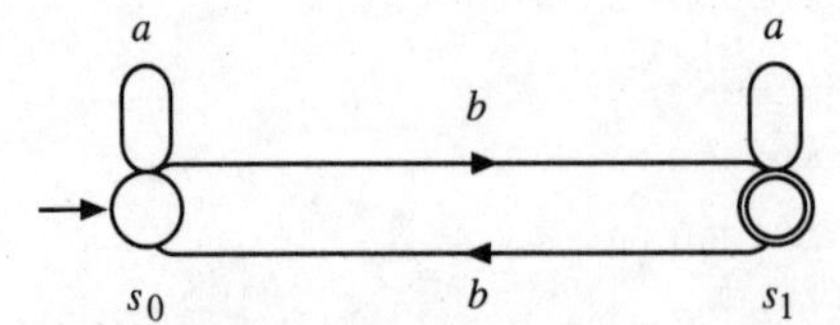

Figure 5.1

Transition tables

We can also describe the behaviour of an FSA by means of a table that shows what transition takes place given any state and input. We label the rows of the table by state, and the columns by input. The accepting states are labelled with asterisks, and the initial state has subscript 0. This way of describing an FSA is more useful for implementing an FSA as part of a program.

Example

5.10 The transition table for the FSA that accepts those strings with an odd number of b's is given in Figure 5.2.

	a	b
s_0	s_0	s_1
s_1^*	s_1	s_0

Figure 5.2

Converting between descriptions Given an FSA described in any of the above ways, you should be able to convert the description to any of the others, *i.e.* given a transition table or a graph, you should be able to construct the other.

Self-Assessment Question 5.3

Discuss the possible advantages of having these different ways of describing an FSA.

Exercise 5.3

1. Construct an FSA to accept strings over $\{a, b\}$ with an even number of b's, and trace the behaviour of your FSA on the two strings *ababab* and *bababba*.

2. Draw the diagram for the FSA that recognizes those strings with an even number of b's. (Remember that 0 is an even number.)

3. Draw up the transition table for the FSA that accepts strings with an even number of b's.

4. (a) Draw the diagram for the FSA whose transition table is given in Figure 5.3.

	a	b	c
s_0	s_1	s_2	s_0
s_1^*	s_0	s_1	s_0
s_2	s_2	s_2	s_2

Figure 5.3

(b) Construct the transition table for the FSA given by Figure 5.4.

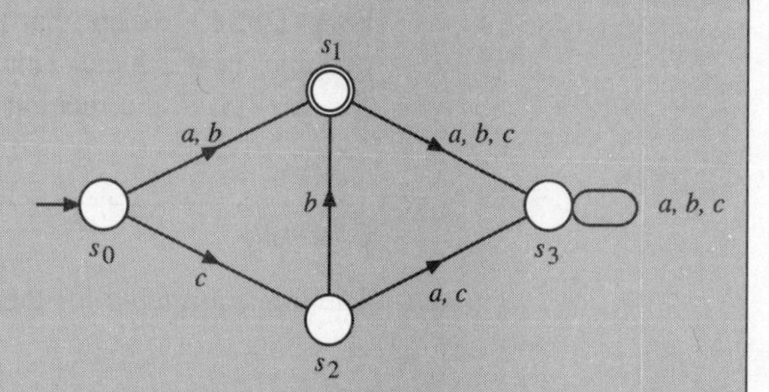

Figure 5.4

(c) Construct the diagram for an FSA that recognizes the language given by a^*ba^*. Also give its transition table, and trace its behaviour on the strings *baa*, *ab*, *aaba* and *aababa*.

5.4 Non-deterministic automata

The automata we have considered so far have had the property that given the current state and input, the next state is completely specified. Sometimes, though, it is convenient to have a choice of next states to go to. In this case, we say that the automaton is **non-deterministic**. If we allow this freedom, it can make it much easier to design an automaton to do some jobs. On the other hand, what does it mean for such an automaton to accept a string? We say that such an automaton accepts a string if there is a possible set of states finishing at an accepting state.

Example

5.11 The non-deterministic finite state automaton (NDFSA) in Figure 5.5 has two possible behaviours on the string 01. It may go to state s_1, then to s_0, or to s_2 and stick there. Since the first choice leaves it in an accepting state, the string is accepted by the automaton.

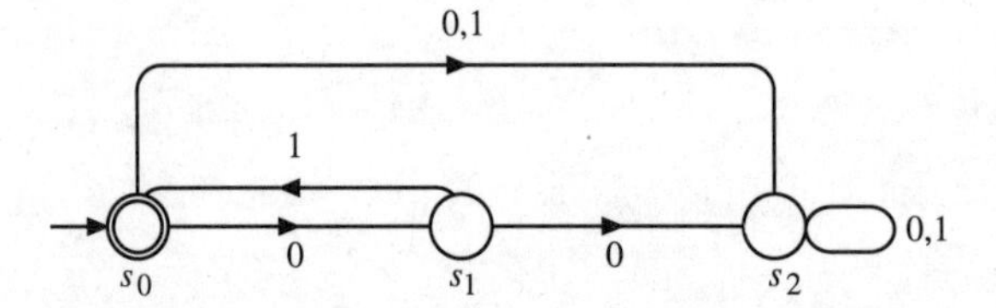

Figure 5.5

The transition table for an NDFSA looks just like that for a deterministic FSA (DFSA), except that rather than listing the next state, it lists the set of possible next states. This means that instead of being a function from $S \times \Sigma \to S$, d is a relation on $(S \times \Sigma) \times S$.

Example

5.12 The transition table for the NDFSA of Figure 5.5 is shown in Figure 5.6.

	0	1
s_0^*	s_1, s_2	s_2
s_1	s_2	s_0
s_2	s_2	s_2

Figure 5.6

Now, you might think that allowing this extra behaviour would make the automata more powerful – in fact this is not so. For any NDFSA there is a DFSA that accepts exactly the same strings. In fact, this equivalent DFSA is quite easy to construct; the basic idea is to think of the NDFSA as simultaneously going into all possible states, and constructing a DFSA that carries this information. It is plausible, then, that the states of our new DFSA must correspond to each of the possible sets of states of the original NDFSA, and the transition table must somehow capture the 'all possibilities' aspect of the NDFSA.

In fact, this is just what we do. Given an NDFSA, M, consisting of a set S of states, and with transition relation d, we construct a DFSA, M', with states S', where the elements of S' are matched up with the (non-empty) subsets of S. The transition function d' is defined by $d'(S_i, a) = S_j$ if there is a state in the set corresponding to S_i related to a state in the set corresponding to S_j by the input a. The initial state of M' is given by the state corresponding to the set containing only the initial state of M. Accepting states are those corresponding to any set of states containing an accepting state.

Worked example

5.13 Construct the DFSA equivalent to the NDFSA of Example 5.12.

Solution First, rather than work with the diagram we work with the transition table. Now, we need our new states, which we list here as $S_0 \sim \{s_0\}$, $S_1 \sim \{s_1\}$, $S_2 \sim \{s_2\}$, $S_3 \sim \{s_0, s_1\}$, $S_4 \sim \{s_0, s_2\}$, $S_5 \sim \{s_1, s_2\}$ and $S_6 \sim \{s_0, s_1, s_2\}$. To see how to construct the transition table for the DFSA, consider $d'(S_4, 0)$. Now, S_4 corresponds to $\{s_0, s_2\}$ and $d(s_0, 0) = \{s_1, s_2\}$, $d(s_2, 0) = s_2$, so that $d'(S_4, 0) \sim \{s_1, s_2\}$, *i.e.* $d'(S_4, 0) = S_5$. In the same way, we obtain the other entries in the transition table of Figure 5.7.

	0	1
S_0^*	S_5	S_2
S_1	S_2	S_0
S_2	S_2	S_2
S_3^*	S_5	S_4
S_4^*	S_5	S_2
S_5	S_2	S_4
S_6	S_5	S_4

Figure 5.7

From this, we can construct the diagram shown in Figure 5.8. Note that not all of the states which appear in the table appear in the diagram. This is because only some of the states can be reached from S_0, and we do not bother to include those states which cannot. For this reason the DFSA constructed has fewer than 8 states. This happens quite often when one constructs DFSA from NDFSA.

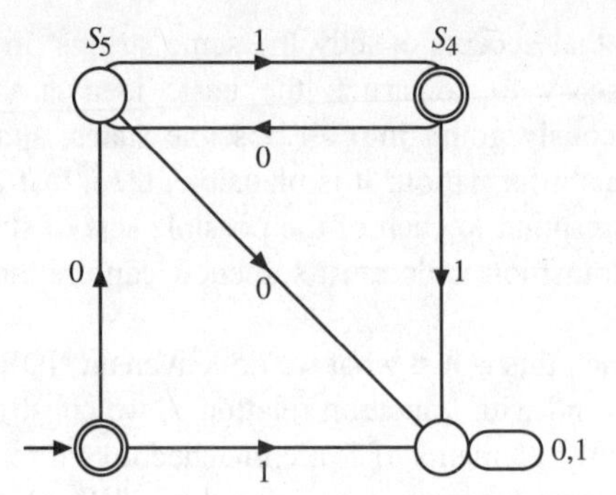

Figure 5.8

Now, you may be wondering why anybody would bother with the non-deterministic machines when we have just seen how to construct a deterministic one to do the same job as any given non-deterministic one. The point is that sometimes it is easy to construct a non-deterministic machine, and then (if necessary) construct an equivalent deterministic one, when it may be difficult to construct a deterministic one in the first place.

For example, suppose we have two machines, M_1 and M_2. M_1 has states $s_{01} \ldots s_{m1}$, transition function d_1, and recognizes language L_1. M_2 has states $s_{02} \ldots s_{n2}$, transition function d_2, and recognizes language L_2. We take the start state of M_1 to be s_{01}, that of M_2 to be s_{02}, and both languages to be over the same alphabet.

We can construct a new machine M to recognize $L_1 \cup L_2$ out of M_1 and M_2. The states of M consist of all states of M_1 and M_2, together with a new start state, s_0. The transitions are given by keeping all the transitions of M_1 and M_2, together with some involving s_0. We take $d(s_0, a) = s_{i1}$ if $d_1(s_{01}, a) = s_{i1}$, and $d(s_0, a) = s_{i2}$ if $d_2(s_{02}, a) = s_{i2}$, for any a in the alphabet. The accepting states of M are all those accepting states from M_1 and M_2, together with s_0 if either language includes the empty string.

Worked examples

5.14 Let L_1 be the language of strings over $\{0, 1\}$ containing an even number of 1's, and L_2 be that of strings containing an even number of 0's. Construct automata for each of these, and also one for the union of the two languages.

Solution The machines that recognize L_1 and L_2 are M_1 and M_2, shown in Figure 5.9, and so the machine that recognizes $L_1 \cup L_2$ is M, shown in Figure 5.10.

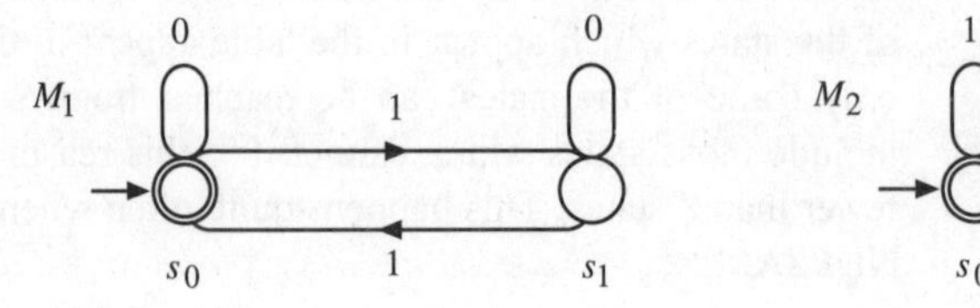

Figure 5.9

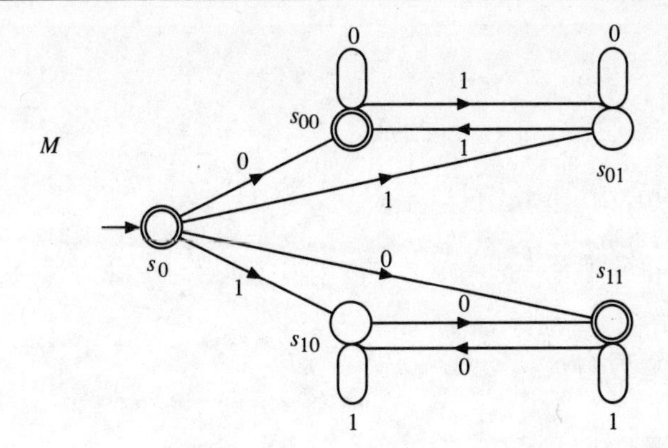

Figure 5.10

We can also construct a machine to recognize the concatenation L_1L_2 of the two languages; this time, we keep all the original states and transitions and include some new transitions, according to the following prescription. If there is a transition leading from any state of M_1 to an accepting state for a given input, include a transition from that state with that input to the start state of M_2.

5.15 Construct a machine that recognizes the language consisting of strings over $\{0, 1\}$ that can be split up into the concatenation of a string with an even number of 1's and one with an even number of 0's.

Solution We build the FSA out of those for recognizing L_1 and L_2 in the previous example, but using the method described just above. The resulting diagram is shown in Figure 5.11.

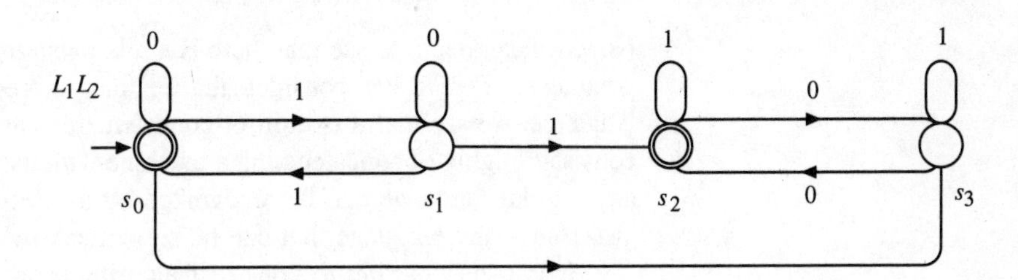

Figure 5.11

Self-Assessment Question 5.4

What is the point of considering NDFSA when anything that can be done by one of them can be done by a DFSA?

Exercise 5.4

1. Which of the following strings are accepted by the machine shown in Figure 5.5: 0101, 11010001, 010101, 010101010101, 1010101010, 011001000011? What is the language accepted by this machine?

2. What is the diagram for the NDFSA with the transition table shown in Figure 5.12?

	a	b
s_0	s_1, s_2	s_0, s_2
s_1^*	s_1, s_0	s_2
s_2	s_0, s_1, s_2	s_1

Figure 5.12

3. Draw the diagram for the NDFSA with the transition table shown in Figure 5.13 and describe the language it recognizes, both in English and by finding a regular expression defining it.

	a	b
s_0	s_1, s_2	s_3
s_1	s_1	s_4
s_2^*	s_4	s_3
s_3^*	s_2	s_4
s_4	s_4	s_4

Figure 5.13

4. Construct the DFSA equivalent to the NDFSA from question 2 in this exercise and draw its diagram.

5. Construct FSA to recognize the languages L_1, defined by the regular expression a^*b^*, and L_2, defined by b^*a^*. Combine these FSA to give FSA that recognize

(a) L_1L_2
(b) $L_1 \cup L_2$

5.5 Relationships between languages and FSA

Now, we can begin to see that there is a relationship between FSA and regular languages. For if we consider the regular expression defining a regular language, we see that it is built of concatenations and unions; the * operation consists roughly of concatenating a machine with itself. It is thus plausible that any regular language can be recognized by an FSA. In fact, the converse is also true – any language that can be recognized by an FSA is regular.

It is even possible to construct algorithms which will convert between regular expressions and FSA; however, they require rather more work and some more theory, which would take us beyond the scope of this book. We will have to settle for accepting the fact, and doing simple examples.

Worked example

5.16 Find a regular expression and an FSA that recognizes strings over $\{0, 1\}$ of even length.

Solution One possible regular expression that does this is $((0 \vee 1)(0 \vee 1))^*$. An FSA that will do the same job is given by the transition table shown in Figure 5.14.

	0	1
s_0^*	s_1	s_1
s_1	s_0	s_0

Figure 5.14

We should be aware, though, that not all languages can be recognized by an FSA. This is because an FSA has strict limitations on the amount of information it can hold, and many languages require more.

Worked example

5.17 Show that the language $L = \{a^n b^n | n \in \mathbb{N}\}$ is not regular.

Solution We prove this by contradiction. So let us assume that L is, in fact, regular, and let M be an FSA that recognizes L. Let the number of states of M be k. Now, consider the string $a^{k+1}b^{k+1}$. By hypothesis, M accepts this string. But while processing the first half, the machine must be in some state, say s_i, more than once (since there are only k states and $(k+1)$ a's in the string). This means that there is a loop in M, starting and finishing at s_i. Suppose the number of states in the loop is p. Then M also accepts the string $a^{k+p+1}b^{k+1}$, since when we get to s_i we can use the next p a's to get back to s_i. Thus M accepts strings not in L, which contradicts the assumption that M recognizes the language L.

Note: This proof strategy is used quite often in the attempt to prove that a given language is not regular. One can state a general result, called the **pumping lemma**, using this form of proof, but for our purposes the idea of the proof is enough. You can find out more about this type of theorem in the books mentioned at the end of this chapter.

Self-Assessment Question 5.5

What use might the relationship between regular expressions and FSA be? (*Hint*: think about editors.)

Exercise 5.5

1. Draw the diagram for the FSA of Worked example 5.16 and see how its structure corresponds to that of the regular expression.
2. Show that the language $\{a^m b^n | m, n \in \mathbb{N} \wedge m < n\}$ is not regular.

5.6 Transducers

An FSA that has been extended by associating to each transition an output is called a transducer. A transducer may be regarded as a kind of translating device or as a simple computer, since it takes input and produces output.

Worked example

5.18 Construct a transducer to calculate the sum of two binary numbers.

Solution Because of the memory limitations of an FSA, we cannot do this by inputting the two numbers separately as strings, then outputting the sum. Instead, we take as input a string consisting of pairs of binary digits, namely the first digit of each number, then the second and so on, and output at each stage the appropriate digit of the sum. For this, all we need to keep track of is how many 1's are coming in, and whether there is a carry digit at the current stage. In the diagram for a transducer, we label each transition by the inputs and the outputs, separating input from output by a colon.

For example, the machine with the diagram shown in Figure 5.15 has state s_0 corresponding to there being no carry, and s_1 corresponding to a carry. It outputs a 1 when it is in state s_0 and the input is either 01 or 10, and so on. Its effect is to produce digit by digit the sum of two binary numbers fed to it digit by digit. This machine will correctly add together two binary numbers, provided that they are the same length and each begins with a 0.

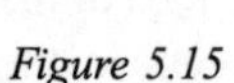
Figure 5.15

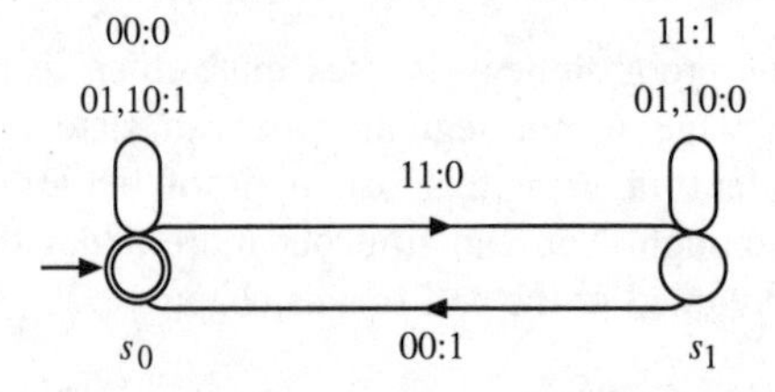

Self-Assessment Question 5.6

Could a transducer be built to evaluate general arithmetic expressions?

Exercise 5.6

Trace the behaviour of the machine from Worked example 5.18 when it is used to add together the binary numbers 011100110 and 010110001.

5.7 A note on applications

Finite state automata are interesting objects in their own right, and many people have invested a great deal of effort in understanding them for their own sake. However, they can also be used as tools. For example, if one wishes to do certain types of pattern matching, an FSA may be the natural way to check a string for conformity to that pattern; and the transition table representation of an FSA is an ideal way of coding the machine in a programming language. A more ambitious aim might be to write a compiler; if the compiler is for a regular language, an FSA to check for syntactic correctness may well be a useful part of the compiler. Even if the whole language is not regular, certain parts of it are likely to be.

Example

5.19 The FSA shown in Figure 5.16 recognizes signed real numbers in exponential notation, where the exponent is restricted to being positive.

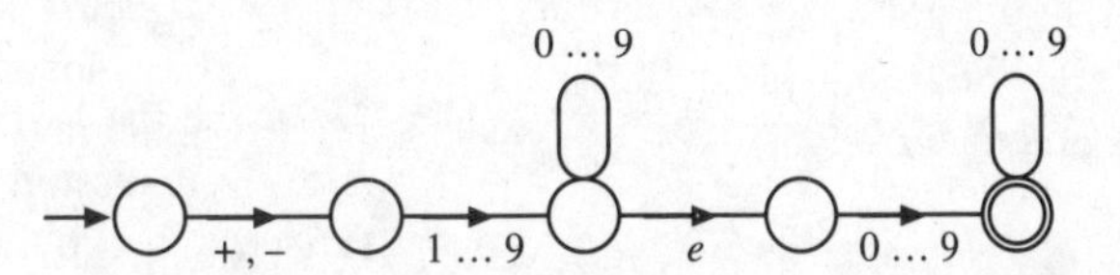

Figure 5.16

Exercise 5.7

Adapt this FSA to recognize numbers where the exponent may be negative as well as positive.

Test and Assignment Exercises 5

1. Let $A = abc$ and $B = cba$.

   (a) Write out AB and BA in full.
   (b) Write $abcabccba$ in terms of A and B.

2. Show that $\#(A^n) = n\#(A)$. (*Hint*: use induction and the fact that $\#(AB) = \#(A) + \#(B)$.)

3. Which of the following are regular expressions?

   (a) $(01)^*(0 \vee 1)^*0$ over $\{0, 1\}$.
   (b) $ab^{**}0$ over $\{a, b, 0\}$.
   (c) $(ab)^* \vee ac$ over $\{a, b\}$.

4. Describe the regular language defined by each of the regular expressions over $\{0, 1\}$.

   (a) $0^*1(0^*10^*10^*)^*0^*$
   (b) $(0 \vee 1)^*$

5. Give a regular expression that defines the language comprising
   (a) strings over $\{a, b\}$ with odd length
   (b) strings over $\{0, 1, 2, 3, 4, 5, 6, 7, 8, 9\}$ ending in 2 and starting with anything other than 0.

6. Consider the DFSA shown in Figure 5.17.

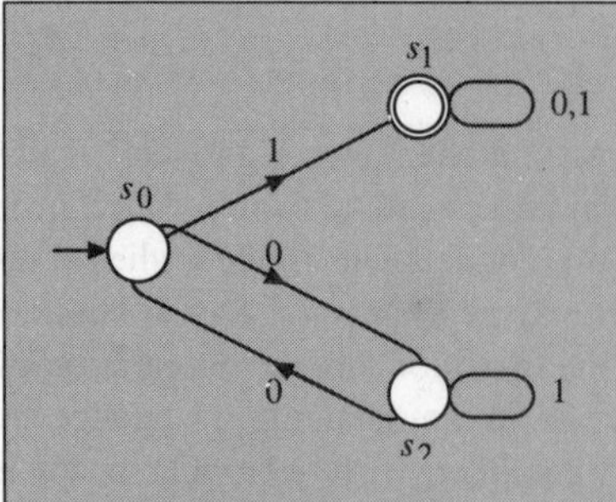

Figure 5.17

(a) Trace its behaviour on each of the strings 000010110, 01001101 and 1001.
(b) What language does this DFSA accept?
(c) Find a regular expression that describes this language.

7. Draw the transition table for the DFSA shown in Figure 5.18. Find a regular expression that describes the language accepted by this automaton.

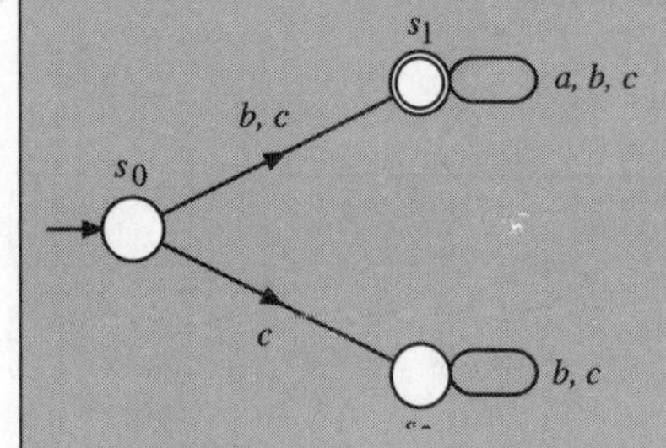

Figure 5.18

8. Draw the directed graph representation of the DFSA with the transition table given in Figure 5.19 and find a regular expression that describes the language recognized by this DFSA.

	a	b
s_0	s_1	s_2
s_1	s_0	s_3
s_2^*	s_2	s_2
s_3	s_2	s_1

Figure 5.19

9. Draw a directed graph that recognizes the language consisting of strings over $\{a, b, c\}$ whose lengths are multiples of three.

10. Give the transition table for the NDFSA whose diagram is shown in Figure 5.20. Find a regular expression describing the language accepted by this automaton.

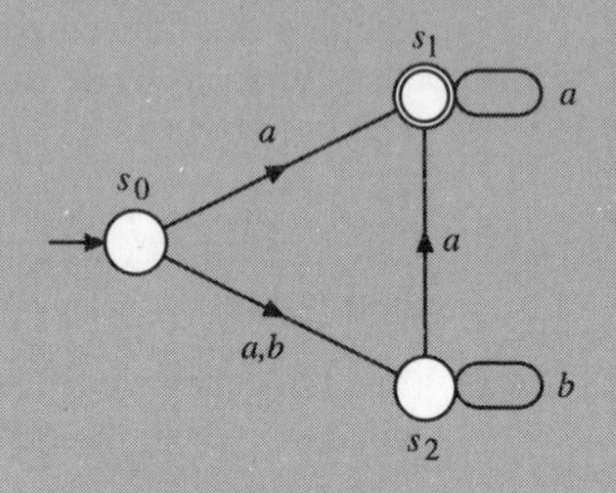

Figure 5.20

11. Let L_1 be the language described by the regular expression $(a \vee b)^*bb$ and L_2 be that described by $(ab \vee ba)^*$. Find automata that recognize L_1, L_2, $L_1 \cup L_2$, L_1L_2 and L_2L_1.

12. Construct a DFSA that recognizes the same language as the NDFSA in Question 10.

13. Construct a DFSA that recognizes the same language as the NDFSA with the table shown in Figure 5.21.

	a	b	c
s_0	s_2	s_2	s_1, s_2
s_1^*	s_3	s_3	s_3
s_2	s_4	s_4	s_1, s_2
s_3	s_3	s_3	s_3

Figure 5.21

14. Construct a transducer that subtracts two binary numbers digit by digit.

5.8 Further reading

J E Hopcroft and J D Ullman, *Introduction to Automata Theory, Languages and Computation* (Addison-Wesley, 1979)
A thorough investigation of languages, machines, computability and the relationships between these concepts.
M Minsky, *Computation: Finite and Infinite Machines* (Prentice-Hall International, 1972)
A detailed study of the various mathematical machines that can be used to model computation.

6 Probability

Objectives

When you have mastered this chapter you should be able to

- use combinatorial ideas to calculate probabilities
- calculate and work with conditional probabilities
- use probability for simple quality control exercises
- apply probabilistic arguments in a variety of situations

Probability theory provides a valuable technique for the analysis of various situations involving uncertainty or risk. We will consider some of the elementary aspects of discrete probability theory, and their applications to a few different areas.

6.1 Introductory probability

A motivational example

Let us begin by considering an example which you can easily test for yourself. If you toss two coins at the same time, what is the probability that they will both show the same, *i.e.* be both heads or both tails?

At first thought, one might reason as follows: either they will both be heads, or one will be heads and the other tails, or both will be tails. Therefore, there is a two out of three chance that both coins will show the same.

But if you try the experiment of tossing two coins many times, you will find that in fact they both show the same only about half the time. They will not show the same exactly half of the time (unless you are very lucky), but if you toss your coins often enough the fraction will be close to $1/2$. What, then, has gone wrong with our analysis?

We can see the problem if we label the coins, say A and B. Doing this, we see that A may come down either heads or tails, as may B, and we can draw up a table of all possibilities,

A	B
heads	heads
heads	tails
tails	heads
tails	tails

so that there are actually four possibilities, of which two correspond to both coins showing the same. If the coins are fair, then each of these possibilities should happen with approximately equal frequency, so that the probability of both coins showing the same is in fact $1/2$, as you would find experimentally.

This little example brings home two lessons: first, probability is related to how often something happens in the long run, and second, we must be very careful in our analysis of just how this something can happen.

Frequency and sample space

We will now put the idea of probability on a slightly more formal footing.

KEY POINT

If an experiment has a number of possible outcomes, we say that the **probability** of some outcome A is $P(A)$, which is defined to be the ratio $N(A)/N$ of $N(A)$ for large N. In other words, $P(A)$ is the ratio of the number of times A is the outcome to N, the number of times the experiment is performed, for large N. As a consequence, if we have good reason for believing that the probability of A is, say, $1/3$, then we expect that if the experiment is carried out many times, the outcome will be A about a third of the time.

We will call an experiment where we do not know the result for certain, but only know the probabilities of the various possible results, a **random experiment**. The possible outcomes of such an experiment are called **elementary events**, and the set of all elementary events is called the **sample space**. Any subset of sample space describes an **event**.

Worked example

6.1 What is the sample space for the experiment of tossing two coins? What subsets of the sample space correspond to the outcomes (i) both show the same, and (ii) the two coins come up different?

Solution If we write H for heads, and T for tails, then the possible outcomes for the experiment are HH, HT, TH and TT, where we imagine that the coins are labelled A and B, and the first item in each pair is what coin A shows and the second is what B shows. Thus the sample space is $\{HH, HT, TH, TT\}$. The event of both coins coming up the same is given by $\{HH, TT\}$, and the event of both coins coming up different is given by $\{HT, TH\}$.

It is traditional to use Ω for the sample space of an experiment; then if $A \subseteq \Omega$ is any event, the probability of A is simply the sum of the probabilities of all the elementary events in A. It is an immediate consequence of this that if A and B describe events and are such that $A \cap B = \emptyset$, then $P(A \cup B) = P(A) + P(B)$. When this is the case, we say that events A and B are **mutually exclusive**. Thus events are mutually exclusive if they cannot both happen. Also, it is always the case that $P(\Omega) = 1$. As far as we are concerned, if $P(A) = 1$, then the event described by A is certain to occur; if $P(A) = 0$, then A never occurs, *i.e.* A is impossible.

Worked examples

6.2 Consider the sample space for the experiment of tossing two coins. Let A be the event 'at least one head shows', B the event 'exactly two heads show', and C the event 'at least one tail shows'.

1. Write out A, B and C.
2. Write out $A \cap C$, and describe the event in English.

Solution
1. $A = \{HH, HT, TH\}$, $B = \{HH\}$, $C = \{HT, TH, TT\}$.
2. $A \cap C = \{HT, TH\}$. This is the event that one coin comes up heads and the other comes up tails.

6.3 Let Ω be some sample space, with subsets A, B and C.

1. What do the sets (i) $A \cap B$, (ii) $A \cup C$, (iii) $A \cap B'$, and (iv) $A \cap (B \cup C)'$ represent?
2. Write down the set-algebraic expressions corresponding to the events (i) 'both B and C happen', (ii) 'A happens but C does not', and (iii) 'A or C happens, but not B'.

Solution
1. (i) A and B both happen, (ii) A or C happens, (iii) A happens but B does not, (iv) A happens, but neither B nor C do.
2. (i) $B \cap C$, (ii) $A \cap C'$, (iii) $(A \cup C) \cap B'$.

Example

6.4 In the experiment of tossing two coins, the events 'both coins come up the same' and 'the coins come up different' are mutually exclusive, since both cannot be the outcome of a single experiment. On the other hand, the events 'both coins come up the same' and 'at least one coin comes up heads' are not mutually exclusive, since *HH* satisfies both conditions.

We also note that if we denote the first event by A and the second by B, then

$$P(A) = 1/2, \ \ P(B) = 3/4, \ \ P(A \cup B) = 3/4$$

so in this case it is certainly not the case that $P(A \cup B) = P(A) + P(B)$.

Clearly, to calculate probabilities of events we need to know the probabilities of the elementary events. These may be known experimentally, or deduced from some more basic properties of the system, or simply assumed as a good working hypothesis which may later be rejected because the probabilites calculated do not correspond well to experimental results.

Worked example

6.5 What is the probability that when two coins are tossed, the result is that the two come up the same?

Solution We have the list of elementary events. Two of the four correspond to the coins coming up the same. Assuming that the coins are fair, the elementary events are all equally likely, so we assign a probability of $1/4$ to each of them. Thus, the probability that the two coins come up the same is $1/4 + 1/4 = 1/2$.

Note that if we have some event A, then A' is just the event that A doesn't happen; since $A \cup A' = \Omega$, we find that $P(A') = 1 - P(A)$. This can be used to simplify some calculations.

Worked example

6.6 What is the probability that when you toss two coins, the coins do not both show heads?

Solution We could work this out by enumerating all the ways the coins could come up other than two heads, but it is simpler to observe that the probability of two heads is $1/4$, so the probability of not getting two heads is $3/4$.

Self-Assessment Question 6.1

Why does the formula

$$P(A \cup B) = P(A) + P(B)$$

not work if the events A and B are not mutually exclusive?

Exercise 6.1

1. Take two coins and toss them 50 times, recording the result. How often are the two the same, and how often are they different?
2. List the possible ways in which two standard (six-sided) dice may fall. What is the probability that the sum of the numbers showing on two dice is 6?
3. What is the sample space for the experiment consisting of throwing two dice? What is the subset of the sample space corresponding to the event that the sum of the results is 5?
4. Write out the subsets of the sample space for the experiment of throwing two dice that describe the events
   (a) a double is thrown,
   (b) the total score is 8,
   (c) the score is at least 9,
   (d) the score is at most 7.
5. In the experiment of throwing two dice, let A be the event that a double is thrown, B that the total score is 8, C that the total score is at least 9, and D that the total score is at most 7.
   (a) What do the sets (i) $A \cap B$, (ii) B', (iii) $C \cup D$, (iv) $A' \cap B$ correspond to?
   (b) Write out the sets corresponding to the events (i) a double is not thrown, (ii) a double is thrown and the total is less than 9, (iii) a double is thrown and the total score is at most seven.
6. What is the probability that when two dice are thrown the sum of the faces uppermost is 6? What is the probability that a double is thrown? Are these events mutually exclusive? What is the probability that the sum of the results is 6 or a double is thrown?
7. What is the probability of getting a double when two dice are thrown? What is the probability of not getting a double?

6.2 Simple combinatorics

The factorial function

We will now look at some mathematics which is useful for analysing certain types of situation, where enumerating all the possibilities becomes difficult. To begin with, we will consider all the possible orders that some set of different objects may be put in.

So, suppose we have n distinct objects; how many ways can we list them? For the first element of our list, we have n choices. This leaves us with $n - 1$

choices for the second element, and for each choice of first element we can have any of the $n - 1$ remaining for the second. Thus we have $n \times (n - 1)$ choices for the first two elements in the list. We now have $n - 2$ choices for the third element, so there are $n \times (n - 1) \times (n - 2)$ choices for the first three elements. As we continue forming the list, the product keeps building up in the same way, until for the second last element we have two choices, and for the last element we have one choice. So, all together, there are $n \times (n - 1) \times (n - 2) \times \ldots \times 2 \times 1$ possible orderings for n objects. This product is called the **factorial** of n, usually denoted $n!$, and pronounced 'n factorial'.

Worked example

6.7 How many ways of listing the five letters a, b, c, d and e are there? If one list is chosen at random, what is the probability that the letter c will occur in the middle?

Solution For five objects, there are $5! = 5.4.3.2.1 = 120$ possible orderings. If c is to lie in the middle, then we have four remaining letters to list, and they can go in any order, so there are $4! = 4.3.2.1 = 24$ possible listings with c in the middle. Hence the probability that a randomly chosen list has c in the middle is $4!/5! = 1/5$.

The factorial function is an extremely useful one in analysing situations involving choices. You should pay particular attention, though, to the reasoning that led to it. This style of reasoning will often enable you to enumerate possibilities when there is no obvious formula in terms of factorials, or when the formula is complicated and hard to remember correctly.

Worked example

6.8 How many anagrams are there of the word *banana*?

Solution The first step is to note that the word *banana* contains six letters, and so there are $6! = 720$ possible orders. However, since some letters are repeated, some of the re-orderings are identical to others. We have to work out how many distinct anagrams there are. There are three letter a's, so there are $3! = 6$ orders that the a's could come in, all leading to the same anagram. There are two letter n's, so there are $2! = 2$ orders that the n's could come in, each leading to the same anagram. All together then, each anagram will appear $2 \times 6 = 12$ times when we give all possible orderings of the letters of *banana*. Thus there are $720/12 = 60$ distinct anagrams of *banana*.

We can extend these ideas a little. If we have a collection of objects, we may form selections of these objects in two distinct ways: after each choice, we may either replace the object in the collection or not. Also, we may care about the order in which the objects are collected or not. For now, we will suppose we have a collection of n distinct objects.

Sampling without replacement

First, let's suppose we want to list r of our objects. Then there are n choices for the first item, $n-1$ for the second, $n-2$ for the third, and so on, until we have $n-r+1$ choices for the rth. Thus there are $n \times (n-1) \times \ldots \times (n-r+1)$ ways of picking r objects from n. This is the same as $n!/(n-r)!$, and is sometimes denoted by ${}^nP_r$.

Worked example

6.9 Nine computers are being reviewed in a consumer magazine, which will publish a 'top three' best buys. How many possible top threes are there?

Solution There are 9 possibilities for first place, 8 for second, and 7 for third, so there are $9 \times 8 \times 7 = 504$ possible top threes. Alternatively, we can calculate this using the fact that there are ${}^9P_3 = 9!/6!$ top threes, and use the factorial function on a calculator to find that there are 504 possible top threes.

Note: By and large, if you are using a pocket calculator, it is best not to calculate factorials explicitly if it is not necessary, as they are frequently large enough to cause arithmetic overflow. It is often more sensible to write out the expression and cancel off as many common factors as you can before evaluating it.

This deals with the case when we are choosing from a collection of objects, and care about the order of the list of selected items. Often, though, we simply want to know how many different collections there are, disregarding order. In this case, the question becomes, given a set with n elements, how many subsets of r elements does it have? We already know that we can find $n!/(n-r)!$ ordered collections of r items. Now, each of these collections can be put in any of $r!$ orders, so in other words, we have counted each collection $r!$ times. So, if we don't care about order, there are $n!/[r!(n-r)!]$ ways of choosing r objects from n. There are a couple of different notations used for this quantity:

$$\binom{n}{r} = {}^nC_r = \frac{n!}{r!(n-r)!}$$

Worked example

6.10 1. A computer shop sells seven different types of PC. I want to buy four, to conduct a comparison between them. How many possible choices do I have?

2. A computer assembly line carries out quality control by checking a random ten samples in each batch of one hundred. The batch is passed if none of the sampled items are defective; otherwise it is rejected. What is the probability that a batch containing ten defective items will be rejected?

Solution 1. I have ${}^{7}C_4$ possible choices, and a simple calculation gives me the answer 35.

2. It would be possible to do this by evaluating the probability that the batch contained at least one defective item; however, the calculation involved is complicated. It is much simpler to find the probability that none of the sampled items are defective. The number of ways of sampling ten items from one hundred is ${}^{100}C_{10}$, and the number of ways of sampling ten items from the ninety sound ones is ${}^{90}C_{10}$. Thus the probability that all the sampled items are sound is

$$\begin{aligned} {}^{90}C_{10}/{}^{100}C_{10} &= \frac{90!}{10!80!} \bigg/ \frac{100!}{10!90!} \\ &= \frac{90!10!90!}{10!80!100!} \\ &= \frac{90!}{80!}\frac{90!}{100!} \\ &= \frac{90 \times 89 \ldots 81}{100 \times 99 \ldots 91} \\ &\approx 0.33 \end{aligned}$$

where the symbol $\approx$ means 'is approximately equal to', and so the probability that the batch will be rejected is approximately $1 - 0.33 = 0.67$.

Sampling with replacement

Now, let us suppose that even after being chosen, an item is eligible to be chosen again; or, to put it another way, we are allowed to make repetitions in our choices.

If we want to choose r items from n, and we care about the order, then we have n choices for the first item, n for the second, and so on, and n for the rth. So there are n^r ways of choosing r items from n if repetitions are allowed.

Worked examples

6.11 How many six-letter strings are there over the alphabet $\{a, b, c, d, e, f, g, h\}$?

Solution There are eight letters to choose from and we have to make six choices, so all together there are $8^6 = 262144$ possibilities.

6.12 If four people are choosing which software package to buy, out of a selection of six, what is the probability that they will all choose different packages?

Solution There are a total of $6^4 = 1296$ ways that they can choose the software, and ${}^6C_4 = 15$ ways that they can all choose different packages. Thus the probability that they will all choose different packages is 15/1296 which is approximately 0.0116, or roughly 1%.

Now, let us suppose, as before, that we are not concerned with the order of the chosen items. This is a more complicated situation, and requires a certain degree of cunning. Again, we suppose we have n objects, and are choosing r from them. If we write a list of the n objects in order, then we can represent each choice by placing the appropriate number of ticks by each object. This can then be encoded as a string, by having a sequence of ticks to say how often the current object is picked, followed by a | to signify that we now move on to the next object. Each string consists of r ticks and $(n-1)$ |'s.

Example

6.13 For example, suppose we choose four letters from the set $\{a, b, c\}$, and the choice is *aacc*. The string describing this is $\surd\surd||\surd\surd$. The string $|\surd|\surd\surd\surd$ corresponds to the choice *bccc*.

Then all the choices of r objects from n correspond to the strings containing r ticks and $n-1$ |'s. In total we have $n+r-1$ symbols in the string, which would give $(n+r-1)!$ possibilities, but $n-1$ of them are |'s and r are ticks, so we divide this by $r!(n-1)!$. Thus if repetition is allowed and we don't care about the order, the number of ways of choosing r objects from n is

$$\frac{(n+r-1)!}{r!(n-1)!} = \binom{n+r-1}{r} = {}^{n+r-1}C_r$$

Worked example

6.14 How many different collections of two computers can I make when there are five kinds to choose from?

Solution I want to find out how many collections of two objects, chosen from five, there are, given that repetition is allowed. The answer is ${}^{5+2-1}C_2 = {}^{6}C_2 = 15$.

Self-Assessment Question 6.2

Why is it generally a good idea to cancel off common factors in expressions such as $126!/124!$?

Exercise 6.2

1. How many ways of listing the first six letters of the alphabet are there? If one such list is chosen at random, what is the probability that a and f are first and last respectively?
2. How many distinct anagrams are there of the word *word*?
3. How many distinct anagrams are there of the word *repetition*?
4. How many ways are there of choosing four objects from seven, assuming that we care about the order?
5. In order to improve team motivation, a team manager decides to publish a merit list of the five best workers in his team. If his team contains eleven members, how many possible merit lists are there?
6. If I have a pool of thirty workers, how many teams of eight can I choose?
7. I have fourteen job descriptions in my in-tray, and time to do seven jobs this week. How many choices do I have in how I spend the week, ignoring the order I do the jobs in?
8. If I have a choice of 4 different types of computer, and laboratory desks numbered 1 to 10, how many different ways can I equip the lab?
9. Four students are out shopping for PC's. There are five models that they can choose between. What is the probability that they all buy the same model?
10. I have six computers, and four software packages to test. If I choose a software package at random for each computer, what is the probability that they will all get different packages?
11. A lab technician is installing software in a lab; he has six packages to choose from, and four computers each of which is to have one of the packages installed on it. He is supposed to put a different package on each machine, but is rather careless, and simply picks a package at random each time. What is the probability that each machine gets a different package?

6.3 Independent and exclusive events

Events are said to be **independent** if $P(A \cap B) = P(A)P(B)$. Two easy ways of spotting when events are independent are when they are the results of separate runs of a given experiment, or the results of unrelated experiments.

Example

6.15

1. Separate throws of a die are independent.
2. The results of tossing a coin and throwing a die are independent.

However, it is not always quite so obvious that events are independent.

Worked example

6.16

1. Consider the experiment of picking a random card from a deck. Are the events 'the result is a spade' and 'the result is a queen' independent?
2. A computer lab contains twelve computers. At any given time, for each computer there is a probability of 0.95 that it is not working. What is the probability that

   (a) all the computers are working?
   (b) none of the computers are working?
   (c) at least one computer is working?
   (d) at least one computer is not working?

Solution

1. There is only one queen of spades, so the probability of getting the queen of spades is $1/52$. One quarter of the cards are spades, so the probability of getting a spade is $1/4$. One thirteenth of the cards are queens, so the probability of getting a queen is $1/13$. Now, $1/4 \times 1/13 = 1/52$, which is the probability of the card being both a spade and a queen, so the two events are independent.
2. First, note that for any computer, the probability of it not working is 0.05, and that the computers are all independent.

   (a) The probability that all are working is $0.95^{12} \approx 0.54$.
   (b) The probability that none are working is $0.05^{12} \approx 2.44 \times 10^{-16}$ (which, practically speaking, is indistinguishable from zero).
   (c) The probability that at least one computer is working is approximately $1 - 2.44 \times 10^{-16}$.
   (d) The probability that at least one computer is not working is approximately $1 - 0.54 = 0.46$.

We already know that in the case that A and B are mutually exclusive, then $P(A \cup B) = P(A) + P(B)$. However, in many cases of interest, the events we consider are not mutually exclusive. In this case we have $P(A \cup B) = P(A) + P(B) - P(A \cap B)$. This formula is analogous to that for the number of elements in the union of two sets, and for just the same reasons; the last term is to cancel out the fact that we have counted the elementary events in $A \cap B$ twice.

Worked examples

6.17 The technician in charge of software installation has put the accounts package on half the machines in the lab, and the spreadsheet on two-thirds of them, at random, and in no case paying any attention to what software is already on the machine. What is the probability that any given machine has

1. both packages on it?
2. at least one package on it?
3. neither package on it?

Solution Since the technician paid no attention to what software was on any machine while installing new software, the presence of the accounts package is independent of that of the spreadsheet. If we denote the presence of the accounts software by A and that of the spreadsheet by S, then $P(A) = 1/2$ and $P(S) = 2/3$.

1. Since A and S are independent, $P(A \cap S) = 1/2 \times 2/3 = 1/3$.
2. The probability that a machine has at least one package on it is $P(A) \cup P(S) = P(A) + P(S) - P(A \cap S) = 1/2 + 2/3 - 1/3 = 5/6$.
3. Finally, the probability that a machine has neither package on it is $P((A \cup S)') = 1 - 5/6 = 1/6$.

6.18 When I am working out my taxes, I have a choice between three spreadsheets and two accounting packages. Over the last three years, I have worked out my annual taxes picking one of the available packages at random each time. What is the probability that

1. I used a spreadsheet each time?
2. I never used a spreadsheet?
3. I used a spreadsheet once and an accounting package twice?

Solution Use a string of S's and A's to denote the packages used; so for example, ASS would represent the event that I used an accounting package in the first year, and a spreadsheet in the other two. In any given year, S is the event that I used a spreadsheet, and A that I used an accounts package. Then $P(A) = 2/5$ and $P(S) = 3/5$ each year.

1. $P(SSS) = (3/5)^3 = 0.216$
2. $P(AAA) = (2/5)^3 = 0.064$
3. The probability that I used a spreadsheet once and an accounts package twice is $P(SAA \cup ASA \cup AAS)$. Since the events here are mutually exclusive, $P(SAA \cup ASA \cup AAS) = P(SAA) + P(ASA) + P(AAS)$. Now, $P(SAA) = 3/5 \times 2/5 \times 2/5 = 0.096$, $P(ASA) = 2/5 \times 3/5 \times 2/5 = 0.096$, $P(AAS) = 2/5 \times 2/5 \times 3/5 = .096$, so $P(AAS \cup ASA \cup SAA) = 3 \times 0.096 = 0.288$.

Self-Assessment Question 6.3

Can two outcomes of an experiment be both independent and exclusive?

Exercise 6.3

1. Out of a list of twenty computers, three people independently pick one at random.

   (a) What is the probability that they all pick the same computer?

   (b) Does adding this probability to the probability that they all pick different computers give the answer 1?

2. A computer lab has four PC's, three mainframe terminals, and two workstations. Whenever I use the lab, I sit down randomly at a desk. Last week I used the lab four times. What is the probability that

   (a) I used a PC each time?
   (b) I used a PC once, and the mainframe three times?
   (c) I used a workstation at least once?

6.4 Conditional probability

Finally, we consider a rather different situation; suppose we want to know the probability of the outcome described by the event A, given that we know that the event B describes the outcome.

Worked example

6.19 Two cards are drawn from a pack, without replacement. What is the probability that the second card is an ace, given that the first one is?

Solution If we know that the first card is an ace, that means that the second is drawn from a pack of fifty-one cards, three of which are aces, so the probability that the second card is an ace given that the first is, is $3/51 = 1/17$.

However, we don't always have to work these probabilities out by hand. There is a useful relationship that can often be used to extract information on **conditional probabilities**, as these probabilities are called. This relationship is known as **Bayes' Theorem**, and we will now spend some time considering it.

KEY POINT

> The probability of the event described by A, given that the event is certainly described by B, is denoted by $P(A|B)$ (read as 'the probability of A given B'), and satisfies the relationship
>
> $$P(A|B) = \frac{P(A \cap B)}{P(B)}.$$

This is Bayes' Theorem, and we can see where it comes from with only a little effort.

If we know that B definitely describes the event in question, then B is the sample space under consideration. $P(A|B)$ is the proportion of the sample space occupied by A, and is hence the proportion of B taken up by $A \cap B$; this is simply $\frac{P(A \cap B)}{P(B)}$.

We can use this to obtain a different way of calculating conditional probabilities from the direct method above.

Worked example

6.20 In a sample of 1000 computers, 600 are owned by educational institutions, while 400 are owned by individuals. Of those owned by educational institutions, 12 are faulty, while of those owned by individuals, 13 are faulty. If one computer is picked at random, what is the probability that it is owned by an individual, given that it is faulty?

Solution Let A describe the event that the computer is owned by an individual, and B the event that it is faulty. Then $P(B) = 0.025$, and $P(A \cap B) = 0.013$. Then $P(A|B) = P(A \cap B)/P(B) = 0.013/0.025 = 0.52$; in other words, if a computer is picked at random and found to be faulty, it will belong to an individual about 52% of the time.

We can also use Bayes' theorem to solve problems that were previously done by less sophisticated methods, such as that in Example 6.19.

Worked example

6.21 Two cards are drawn from a pack, without replacement. What is the probability that the second card is an ace, given that the first one is?

Solution Let F describe the event that the first card is an ace, and S the event that the second card is. Then $P(S|F) = P(S \cap F)/P(F)$. Now, $P(S \cap F)$ is the probability of drawing two aces from a pack, which is given by

$$
\begin{aligned}
{}^4C_2/{}^{52}C_2 &= \frac{4!}{2!2!} \Big/ \frac{52!}{50!2!} \\
&= \frac{4!2!50!}{52!2!2!} \\
&= \frac{4 \times 3}{52 \times 51} \\
&= \frac{1}{13 \times 17}
\end{aligned}
$$

And $P(F) = 1/13$, so that $P(S|F) = 1/17$.

Note that in this case, the problem is harder to solve using Bayes' theorem, rather than easier!

More interestingly, we can use conditional probabilities to argue backwards; given $P(A|B)$ and some other items of information, it may be possible to find the perhaps more useful or interesting $P(B|A)$. The two are not generally equal. The relationships that enable us to do this are

$$P(A|B) = \frac{P(A \cap B)}{P(B)}$$

and

$$P(B|A) = \frac{P(A \cap B)}{P(A)},$$

from which we immediately see that

$$P(B|A) = \frac{P(B)}{P(A)} P(A|B).$$

Worked example

6.22 I have in my office two computers, previously used by Alan and Basil. I know that Alan's machine—machine A—contains fourteen FORTRAN programs and six PASCAL programs, while Basil's machine—machine B—contains four FORTRAN programs and sixteen PASCAL programs. Unfortunately, both Alan and Basil had uninformative names for their programs, and both machines contain files called `prog1` to `prog20`. In an attempt to find out which machine is which, I pick one at random, and print out a file. It turns out to be a FORTRAN program. What is the probability, then, that the machine is Alan's?

Solution I will use A to represent the event that the machine is Alan's, and F the event that the file printed out is in FORTRAN.

I need to work out $P(A|F)$. Now, this is difficult to get at directly, but I do know that $P(A|F) = \frac{P(A)}{P(F)} P(F|A)$. Now, $P(F|A) = 14/20 = 7/10$, since there are twenty programs on Alan's machine, of which fourteen are in FORTRAN. The total proportion of FORTRAN programs is eighteen out of forty, so that $P(F) = 18/40 = 9/20$. Finally, $P(A) = 1/2$ since I picked a machine at random. This gives me

$$\begin{aligned} P(A|F) &= \frac{P(A)}{P(F)} P(F|A) \\ &= \frac{10}{9} \times \frac{7}{10} \\ &= \frac{7}{9} \end{aligned}$$

Thus, the probability that I chose Alan's machine is $7/9$.

There are a few important properties of conditional probabilities that can make calculation easier, *viz.*

1. A and B are mutually exclusive if and only if $P(A|B) = 0$.
2. A and B are independent, if and only if $P(A|B) = P(A)$.
3. If A and B are mutually exclusive, then for any event X, we have $P(A \cup B|X) = P(A|X) + P(B|X)$.
4. If A and B are mutually exclusive, and $A \cup B = \Omega$, then for any event X, $P(X) = P(X|A)P(A) + P(X|B)P(B)$.

Worked example

6.23 Owing to a design fault, one in fifty of the disk drives produced by Bodger & Son have an intermittent fault. The resident disk drive expert at Bodger & Son has devised a test for disk drives that is 80% reliable, in the sense that 80% of faulty drives get a positive result on the test, while 80% of good drives get a negative one. What is the probability that a drive that gets a positive result on the test is in fact faulty?

Solution Let us use p and n respectively to describe postive and negative results from the test, and F and G to mean that the drives are faulty and good respectively. Note that $p = n'$ and $F = G'$.

We want to know $P(F|p)$, given that $P(p|F) = P(n|G) = 0.8$, and $P(F) = 0.02$. The other piece of information we require to find $P(F|p)$ is $P(p)$.

Now,

$$\begin{aligned} P(p) &= P(p|F)P(F) + P(p|G)P(G) \\ &= 0.8 \times 0.02 + 0.2 \times 0.98 \\ &= 0.212 \end{aligned}$$

Thus

$$P(F|p) = \frac{P(F)}{P(p)} P(p|F)$$
$$= \frac{0.02}{0.212} \times 0.8$$
$$\approx 0.075$$

Thus approximately 8% of those disk drives that get a positive result in the test are actually faulty. This conflicts rather strongly with most people's intuitive feeling that about 80% of disk drives getting a positive result are actually faulty, and demonstrates the need for accurate analysis when probabilistic results are under consideration.

Self-Assessment Question 6.4

When would you expect the equation $P(A|B) = P(B|A)$ to hold?

Exercise 6.4

1. A programming team contains seven men and three women, each of whom buys one ticket for the team raffle. What is the probability that the second prize goes to a woman, given that the first prize does?
2. George, the team leader, has two programmers working for him, Alberta and Barbara. Three-quarters of Alberta's work is in MODULA-2 and the rest in C++, whereas four-fifths of Barbara's work is in C++ and the rest in MODULA-2. After a team meeting to which Alberta has brought twice as many listings as Barbara, George notices that one of them has left behind a listing, but doesn't know which of them it was. Given that the listing was in C++, what is the probability that it was Barbara's?
3. In the situation described in Worked example 6.23, what is the probability that a disk drive that gets a negative result on the test is actually good?

6.5 Stochastic processes and probability trees

Many problems of interest involve systems which change with time in an unpredictable manner. The case we are interested in here is that of systems where, if you know the current state, you don't know what the state will be when the system is next observed, but you do know the probability of the system moving from its current state to any other. This situation, where we have a sequence of outcomes related by probabilities, is called a **stochastic process**. We are only going to consider the case where the probabilities are fixed, and cannot vary with time.

Example

6.24 Let us consider what the weather is like on successive days. After keeping records for some time, we find that if the weather is good one day, it is good the next day some 70% of the time, and bad the other 30%. On the other hand, if it is bad one day, it is bad the next day 80% of the time, and good the other 20% of the time. So the probability of the weather remaining good tomorrow if it is good today is 0.7, and so on.

There are various ways of analysing stochastic processes, of varying sophistication and difficulty. We will only consider a very simple one, which can be used to analyse a stochastic process with few possible states, over a small number of time steps. The tool we will use is called a **probability tree**, and is a diagram showing all possible histories for the system over the time in question and labelled with the various probabilities in a way that makes it easy to do calculations.

To draw up a probability tree, we start off with an initial point, called a **root** (which may or may not be labelled by a state), and draw lines from there to a collection of points, called **daughter nodes**, and labelled by the subsequent state; each of these lines is labelled by the probability of that change. We then simply repeat the process of joining each state to the set of points labelled by the states which may be reached from there, and labelling the line by the appropriate probability. Then each branch of the tree, starting off at the initial point and ending at a node with no daughters, a **final node**, represents a possible history of the system over the time in question. To find the probability of the system having that history, we simply multiply together the probabilities of each of the transitions involved. To find the probability that the system ends up in some particular state, we just add together the probabilities of each of the possible histories that end up with the system in that state (since the various histories are clearly mutually exclusive).

Worked examples

6.25 Consider the weather system again. Draw up the probability tree for the two days after a fine day, and find the probability that

1. on the last day considered the weather is good,
2. there is at least one day of bad weather in the period considered.

Solution First, we draw up the probability tree; G is used to label days of good weather, and B days of bad weather. Now we can use this tree to calculate the required probabilities.

1. There are two final nodes corresponding to good weather; the sequence of days that lead to them are GBG and GGG, with probabilities $0.3 \times 0.2 = 0.06$ and $0.7 \times 0.7 = 0.49$ respectively. Hence the probability of good weather on the third day is $0.06 + 0.49 = 0.55$.

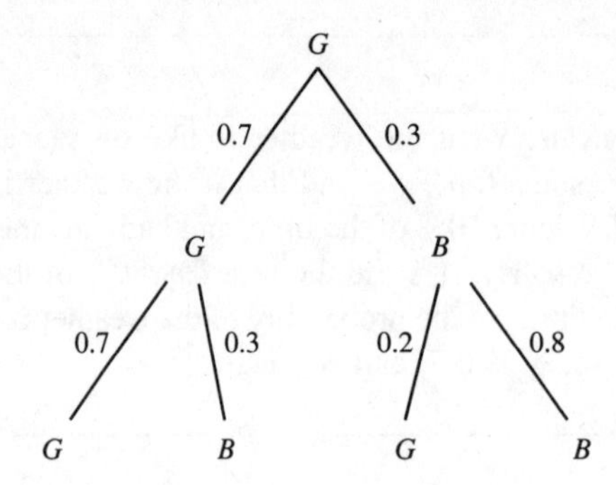

2. Three of the histories correspond to at least one day of bad weather, namely *GGB*, *GBG* and *GBB*. These have associated probabilities of $0.7 \times 0.3 = 0.21$, $0.3 \times 0.2 = 0.06$, and $0.3 \times 0.8 = 0.24$. Hence the probability is $0.21 + 0.06 + 0.24 = 0.51$.

6.26 Draw up the probability tree for tossing a coin twice, and use it to find the probability of getting the same result on each toss.

Solution Again, we begin by drawing up the tree; this time, the initial node has no label, since it corresponds to the time just before we toss the coin for the first time. The daughters are labelled *H* and *T* for heads and tails, and each probability is 0.5.

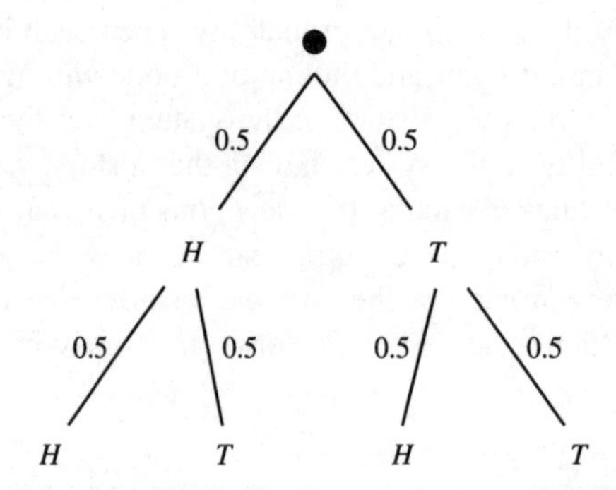

From examination of the diagram we see that there are two possible ways in which both tosses have the same result, namely *HH* and *TT* each of which has a probability of $0.5 \times 0.5 = 0.25$, so the probability is $0.25 + 0.25 = 0.5$.

Self-Assessment Question 6.5

What is the use of probability trees?

Exercise 6.5

We examine the loading on our network server every four hours from 8:00 am, when it becomes available, to 8:00 pm, when it becomes unavailable. We find that on average, if the system is busy when examined, there is a probability of 0.6 that it will be busy when it is next examined, and if it is quiet when examined, there is a probability of 0.5 that it will be quiet when next examined. Yesterday, the system was quiet at its first examination, at 8:00 am. Draw up a probability tree for the state throughout the day, and find the probability that

1. the machine was quiet at 8:00 pm,
2. the machine was busy all day,
3. the machine was busy on two examinations,
4. the machine was busy on at least two examinations.

Test and Assignment Exercises 6

1. Give the sample space for each of the following experiments.
   (a) Three coins are tossed.
   (b) (i) A coin is tossed. (ii) A die is rolled. (iii) A coin is tossed and a die is rolled. What is the relationship between these three?
   (c) The fastest two computers are chosen from a group of four.
   (d) A bag contains five balls, coloured red, white, blue, purple and green. Two are taken out of the bag.
   (e) A bag contains five balls, coloured red, white, blue, purple and green. A ball is taken out of the bag, then replaced, then a ball is taken out.
   (f) A bag contains three balls, two of which are black and one of which is white. (i) One ball is removed. (ii) Two balls are removed. (iii) A ball is removed, then replaced, then a ball is removed.
2. (a) How many ways can I set up a laboratory with desks numbered from 1 to 20 with a selection of PCs and Macintoshes?
   (b) Ten letters are to be placed into ten envelopes. What is the probability that they each go in the correct envelope if they are put in at random?
   (c) A large container holds fourteen computers and twenty boxes of floppy disks. How many ways can I choose three computers and five boxes of floppy disks from this container?
   (d) Seven numerical packages are tested on a series of benchmark problems. How many possible top threes are there if (i) you list them in order of merit, (ii) you list them in no particular order?
   (e) I have six envelopes in front of me. As I am on several mailing lists, four contain complimentary tickets for a computer show, and the other two contain advertisements for the show. How many distinct collections of tickets and advertisements can I make by opening (i) three envelopes? (ii) four envelopes?
3. (a) What is the probability of tossing a coin four times in a row and getting (i) four heads, (ii) two heads and two tails, (iii) three tails and one head?
   (b) In a raffle, you win a computer and a software package. The computer is either a PC or a Macintosh. There are four software packages, three of which will run only on the Mac, and one of which will run only on the PC. What is the probability that you have won a compatible pair?
   (c) A bag contains three balls, one of which is red and two of which are blue. You take out a ball, replace it, and take out a ball. What is the probability that (i) you got the same colour each time? (ii) you got blue both times?
   (d) A room contains three PCs, one of which is made by the company Blue Stripes Ltd, and two of which are

inexpensive imitations made by PC Clones Inc. On two successive days you visit the room to work on a program. What is the probability that both visits involved (i) the use of computers made by the same manufacturer, (ii) the use of the machine built by Blue Stripes Ltd?

(e) If two dice are thrown, what is the probability of (i) a double, (ii) a total of 11, (iii) double 2, (iv) a total of 7?

(f) If three dice are thrown, what is the probability of getting (i) all three the same, (ii) a double?

(g) Four software packages are each to be installed on one of four machines. Unfortunately, the instructions saying which package should go on which machine have been lost, so they are installed at random, one on each machine. What is the probability that they have been installed correctly?

4. (a) Four people are asked to choose their favourite computer from a list of two IBM models, three PC Clones Ltd models and a specialist games machine. What is the probability that (i) they all choose IBM, (ii) none choose IBM, (iii) at least one chooses IBM, (iv) they all choose the same manufacturer, (v) they all choose different manufacturers, (vi) all choose the games machine?

(b) From a list of machine specifications I take out two at random. If seven of the specifications include colour monitors and the remaining four include monochrome, what is the probability that I have picked (i) two colour specifications, (ii) one colour and one monochrome, and (iii) two monochrome?

5. (a) I have to maintain a computer lab containing fourteen computers and seven printers. Each day I pick one device at random to carry out routine maintenance on, making sure only that I don't pick the same item on two successive days. What is the probability that (i) I work on a computer tomorrow, given that I worked on a printer today, (ii) I worked on a printer yesterday, given that I worked on a printer today?

(b) In a certain installation, fifty percent of the machines that have a coprocessor have colour monitors, while only ten percent of the machines without have. If thirty percent of all the machines have coprocessors, what percentage of machines with a colour monitor have a coprocessor? What percentage of all machines in the installation have a coprocessor?

6. (a) In a computer laboratory which I use there are six PCs and three mainframe terminals, any of which I can use to work out my current project. When I arrive, I pick a machine at random. Draw up a probability tree to describe the possibilities for which machines I use on Monday to Thursday, given that I use a PC on Monday. What is the probability that I (i) use the same machine every day? (ii) use a mainframe terminal at least once? (iii) use a mainframe terminal exactly once? (iv) use a PC on Thursday?

(b) A statistically minded user of one of our labs observed that if a user worked with a DOS based machine on one occasion, there was a 40% chance they would also use it the next occasion, and a 60% chance that they would switch to some form of GUI. One the other hand, if a user worked with a GUI on any given occasion, there was a 70% chance they would use a GUI the next time, and a 30% chance they would switch to DOS. John and Bev both used a computer in that laboratory once each day on Monday to Thursday. Draw up a probability tree for each of them given that John used a DOS interface on Monday and Bev used a GUI.

What is the probability that (i) John used a GUI on Thursday? (ii) Bev used DOS on Thursday? (iii) John and Bev used the same interface on Thursday? (iv) John use a DOS interface an even number of times? (v) Bev used a GUI an odd number of times?

6.6 Further reading

W Feller, *An Introduction to Probability Theory and its Applications* (Wiley,1968). A standard text that develops at length the ideas introduced here.
Y A Rozanov, *Probability Theory: A Concise Course* (Dover Publications, 1977). A short text that provides an overview of probability theory.

7 Matrices

Objectives

When you have mastered this chapter you should be able to

- use matrices to represent data
- add, subtract and multiply matrices
- understand what is meant by the inverse of a matrix and be able to calculate it
- represent and solve systems of equations using matrices

This chapter introduces a rectangular arrangement of data values known as a matrix. Matrices have a wide variety of uses in information technology and computing. They provide a means for representing data when using linear programming to solve optimization problems, in the field of data structures they appear as two-dimensional arrays and in computer graphics they can be used to represent transformations, as we shall see in Chapter 8.

7.1 Introducing matrices

In order to introduce matrices and matrix operations we will begin by considering the use of matrices in a simplified and somewhat artificial context.

A small manufacturing company makes just one product which is a specially engineered bolt. The bolt comes in just three sizes, small, medium and large. One of the tasks of the company's sales manager Jim is to record details of the sales of this product. For the purpose of recording these statistics Jim divides the country into three regions: North, West and South. He then records the quantity sold of each size of bolt in each region using tables in the following way:

Quantity of sales in June

	North	West	South
Small	100	200	70
Medium	60	60	90
Large	120	90	110

Quantity of sales in July

	North	West	South
Small	150	210	85
Medium	80	60	100
Large	130	90	120

This process of recording sales statistics makes use of rectangular arrangements of values. In mathematics such a rectangular arrangement of values is known as a **matrix**. For example, the arrangement of values

$$\begin{bmatrix} 100 & 210 & 85 \\ 60 & 60 & 90 \\ 120 & 90 & 110 \end{bmatrix}$$

which has three **rows** and three **columns** is known as a three by three (written as 3×3) **square matrix**.

Matrices come in a variety of shapes and sizes; they may be square, having the same number of rows as columns as in the previous example, or rectangular. The following matrix which has two rows and four columns is an example of a 2×4 rectangular matrix of real numbers:

$$\begin{bmatrix} 2.4 & 3.74 & 0.6 & 2.75 \\ 0.25 & 2.4 & 6.45 & 10.4 \end{bmatrix}$$

A matrix may consist of a single column such as

$$\begin{bmatrix} 3.5 \\ 2.7 \\ 4.5 \\ 0.3 \end{bmatrix} \quad \text{or as a single row such as} \quad \begin{bmatrix} 2.7 & 3.5 & 6.7 \end{bmatrix}$$

In general a matrix which has m rows and n columns is known as an $m \times n$ matrix.

KEY POINT

In this chapter we will use bold capital letters such as $\boldsymbol{A}$, $\boldsymbol{B}$ and $\boldsymbol{C}$ to stand for matrices. A general matrix $\boldsymbol{A}$ which has m rows and n columns may be represented in the following way,

$$\boldsymbol{A} = \begin{bmatrix} a_{11} & a_{12} & a_{13} & \cdots & a_{1n} \\ a_{21} & a_{22} & a_{23} & \cdots & a_{2n} \\ \vdots & \vdots & \vdots & & \vdots \\ a_{m1} & a_{m2} & a_{m3} & \cdots & a_{mn} \end{bmatrix}$$

with each a_{ij} representing a particular data value, the i indicating the row and the j indicating the column in which the value is located. So the data item a_{23} is stored in row 2, column 3.

Self-Assessment Question 7.1

What would be the dimensions of a matrix which consisted of a single column but held 4 data values and what would its dimensions be if it consisted of a single row but held 6 data values?

Exercise 7.1

1. The company described in this section employs four sales representatives who may choose to work in some or all of the regions. During the month of May Jill chooses to work all three regions making sales worth £200, £150 and £400 in the North, West and South respectively. During the same period Jack makes sales worth £200 in the North and £300 in the South, while Jane sells £500 worth of goods in the West and £310 worth in the South. John who works only in the South makes sales worth £600. Represent this data in the form similar to that used in the example above.

2. What are the dimensions of the following matrices?

   (a) $\boldsymbol{A} = \begin{bmatrix} 2 & 6 & 12 \\ 14 & 0 & 9 \\ 10 & 8 & 6 \end{bmatrix}$

   (b) $\boldsymbol{B} = \begin{bmatrix} 2.5 & 4.2 \\ 3.6 & 5.1 \\ 4.2 & 8.4 \end{bmatrix}$

   (c) $\boldsymbol{C} = \begin{bmatrix} \sqrt{2} & 0 & \sqrt{3}/2 \\ 1 & \sqrt{2}/2 & 1 \end{bmatrix}$

   (d) $\boldsymbol{D} = \begin{bmatrix} 6.0 & 7.0 & 8.0 & 4.0 \\ 2.4 & 4.0 & 3.0 & 6.2 \\ 3.2 & 6.5 & 4.0 & 2.0 \end{bmatrix}$

3. With reference to the matrices $\boldsymbol{A}$, $\boldsymbol{B}$, $\boldsymbol{C}$ and $\boldsymbol{D}$ in the previous question write down the values of the following elements of these matrices.

   (a) a_{23} (b) b_{32} (c) c_{11} (d) d_{24}

7.2 Matrix addition and subtraction

In this section we will see how to perform the arithmetic operations of addition and subtraction on matrices.

Matrix addition

Referring back to Section 7.1, let us suppose that Jim is requested to produce accumulated sales figures for June and July. In order to do this he performs the following calculation.

Dropping the row and column labels from the matrices that hold the sales statistics, he writes

$$\begin{array}{cc} \text{Sales in June} & \text{Sales in July} \\ \begin{bmatrix} 100 & 200 & 70 \\ 60 & 60 & 90 \\ 120 & 90 & 110 \end{bmatrix} + & \begin{bmatrix} 150 & 210 & 85 \\ 80 & 60 & 100 \\ 130 & 90 & 120 \end{bmatrix} \end{array}$$

He then combines the two matrices to give

$$\text{Sales in June} + \text{Sales in July}$$
$$\begin{bmatrix} 100+150 & 200+210 & 70+85 \\ 60+80 & 60+60 & 90+100 \\ 120+130 & 90+90 & 110+120 \end{bmatrix}$$

Carrying out the additions he gets

$$\text{Sales in June} + \text{Sales in July}$$
$$\begin{array}{r} \\ \text{Small} \\ \text{Medium} \\ \text{Large} \end{array} \begin{array}{c} \begin{array}{ccc} \text{North} & \text{West} & \text{South} \end{array} \\ \begin{bmatrix} 250 & 410 & 155 \\ 140 & 120 & 190 \\ 250 & 180 & 230 \end{bmatrix} \end{array}$$

We can see that all Jim has done in order to calculate the accumulated sales for June and July is to add together the corresponding entries in the sales matrices for the two months.

KEY POINT

In general two matrices may be added together as long as they have the same dimensions. If we have the matrices $\boldsymbol{A}$ and $\boldsymbol{B}$ of the form

$$\boldsymbol{A} = \begin{bmatrix} a_{11} & a_{12} & a_{13} & \cdots & a_{1n} \\ a_{21} & a_{22} & a_{23} & \cdots & a_{2n} \\ \vdots & \vdots & \vdots & & \vdots \\ a_{m1} & a_{m2} & a_{m3} & \cdots & a_{mn} \end{bmatrix}$$

and

$$\boldsymbol{B} = \begin{bmatrix} b_{11} & b_{12} & b_{13} & \cdots & b_{1n} \\ b_{21} & b_{22} & b_{23} & \cdots & b_{2n} \\ \vdots & \vdots & \vdots & & \vdots \\ b_{m1} & b_{m2} & b_{m3} & \cdots & b_{mn} \end{bmatrix}$$

then the sum of these two matrices is given by

$$\boldsymbol{A}+\boldsymbol{B} = \begin{bmatrix} a_{11}+b_{11} & a_{12}+b_{12} & a_{13}+b_{13} & \cdots & a_{1n}+b_{1n} \\ a_{21}+b_{21} & a_{22}+b_{22} & a_{23}+b_{23} & \cdots & a_{2n}+b_{2n} \\ \vdots & \vdots & \vdots & & \vdots \\ a_{m1}+b_{m1} & a_{m2}+b_{m2} & a_{m3}+b_{m3} & \cdots & a_{mn}+b_{mn} \end{bmatrix}$$

So we see that the addition of two matrices is achieved by adding together the data values that lie in corresponding positions.

Worked example

7.1 With reference to the matrices from Exercise 7.1 and

$$\boldsymbol{E} = \begin{bmatrix} 2.5 & 6.5 \\ 4.2 & 3.2 \\ 2.0 & 1.5 \end{bmatrix} \qquad \boldsymbol{F} = \begin{bmatrix} 2 & 0 & 4 \\ 1 & 1 & 2 \\ 3 & 2 & 4 \end{bmatrix} \qquad \boldsymbol{G} = \begin{bmatrix} a & b \\ c & d \end{bmatrix}$$

$$\boldsymbol{H} = \begin{bmatrix} 2a & 2b+c \\ c+d & d^2 \end{bmatrix} \qquad \boldsymbol{J} = \begin{bmatrix} x \\ y \\ z \end{bmatrix} \qquad \boldsymbol{K} = [2.5 \quad 3.6 \quad 4.2]$$

$$\boldsymbol{L} = \begin{bmatrix} x+1 \\ y^2 \\ 3z \end{bmatrix} \qquad \boldsymbol{M} = \begin{bmatrix} \sqrt{2} & 6 \\ 4 & 8 \end{bmatrix}$$

where possible perform the following calculations.

(a) $\boldsymbol{B}+\boldsymbol{E}$ (b) $\boldsymbol{E}+\boldsymbol{B}$ (c) $\boldsymbol{G}+\boldsymbol{H}$ (d) $\boldsymbol{A}+\boldsymbol{D}$ (e) $\boldsymbol{G}+\boldsymbol{M}$

Solution (a) $\begin{bmatrix} 5.0 & 10.7 \\ 7.8 & 8.3 \\ 6.2 & 9.9 \end{bmatrix}$ (b) $\begin{bmatrix} 5.0 & 10.7 \\ 7.8 & 8.3 \\ 6.2 & 9.9 \end{bmatrix}$ (c) $\begin{bmatrix} 3a & 3b+c \\ 2c+d & d+d^2 \end{bmatrix}$

(d) Cannot be done since the matrices $\boldsymbol{A}$ and $\boldsymbol{D}$ have different dimensions.

(e) $\begin{bmatrix} a+\sqrt{2} & b+6 \\ c+4 & d+8 \end{bmatrix}$

Matrix subtraction

Referring back to Section 7.1, suppose the sales manager is asked to produce statistics showing the change in the quantity of sales of bolts between the months of June and July. To do this he could perform the following calculation:

Sales in July – Sales in June

Writing this in terms of the sales matrices we have

$$\overset{\text{Sales in July}}{\begin{bmatrix} 150 & 210 & 85 \\ 80 & 60 & 100 \\ 130 & 90 & 120 \end{bmatrix}} - \overset{\text{Sales in June}}{\begin{bmatrix} 100 & 200 & 70 \\ 60 & 60 & 90 \\ 120 & 90 & 110 \end{bmatrix}}$$

giving

Sales in July – Sales in June

$$\begin{bmatrix} 150-100 & 210-200 & 85-70 \\ 80-60 & 60-60 & 100-90 \\ 130-120 & 90-90 & 120-110 \end{bmatrix}$$

Carrying out the subtractions gives

Sales in July – Sales in June

$$\begin{bmatrix} 50 & 10 & 15 \\ 20 & 0 & 10 \\ 10 & 0 & 10 \end{bmatrix}$$

Just as with addition we see that the process of subtraction has taken place between corresponding elements of the two matrices. So in general if one matrix is to be subtracted from another then the two matrices must have the same dimensions.

KEY POINT

Using the general matrices $\boldsymbol{A}$ and $\boldsymbol{B}$ we may represent the general case for subtraction in the following way:
With

$$\boldsymbol{A} = \begin{bmatrix} a_{11} & a_{12} & a_{13} & \cdots & a_{1n} \\ a_{21} & a_{22} & a_{23} & \cdots & a_{2n} \\ \vdots & \vdots & \vdots & & \vdots \\ a_{m1} & a_{m2} & a_{m3} & \cdots & a_{mn} \end{bmatrix}$$

and

$$\boldsymbol{B} = \begin{bmatrix} b_{11} & b_{12} & b_{13} & \cdots & b_{1n} \\ b_{21} & b_{22} & b_{23} & \cdots & b_{2n} \\ \vdots & \vdots & \vdots & & \vdots \\ b_{m1} & b_{m2} & b_{m3} & \cdots & b_{mn} \end{bmatrix}$$

$$\boldsymbol{A} - \boldsymbol{B} = \begin{bmatrix} a_{11} - b_{11} & a_{12} - b_{12} & a_{13} - b_{13} & \cdots & a_{1n} - b_{1n} \\ a_{21} - b_{21} & a_{22} - b_{22} & a_{23} - b_{23} & \cdots & a_{2n} - b_{2n} \\ \vdots & \vdots & \vdots & & \vdots \\ a_{m1} - b_{m1} & a_{m2} - b_{m2} & a_{m3} - b_{m3} & \cdots & a_{mn} - b_{mn} \end{bmatrix}$$

Worked example

7.2 With reference to the matrices in Exercise 7.1 and Worked example 7.1, where possible, perform the following calculations.

(a) $\boldsymbol{B} - \boldsymbol{E}$ (b) $\boldsymbol{E} - \boldsymbol{B}$ (c) $\boldsymbol{F} - \boldsymbol{A}$ (d) $\boldsymbol{J} - \boldsymbol{L}$ (e) $\boldsymbol{A} - \boldsymbol{E}$ (f) $\boldsymbol{H} - \boldsymbol{G}$

Solution

(a) $$\begin{bmatrix} 2.5 & 4.2 \\ 3.6 & 5.1 \\ 4.2 & 8.4 \end{bmatrix} - \begin{bmatrix} 2.5 & 6.5 \\ 4.2 & 3.2 \\ 2.0 & 1.5 \end{bmatrix} = \begin{bmatrix} 0.0 & -2.3 \\ -0.6 & 1.9 \\ 2.2 & 6.9 \end{bmatrix}$$

(b) $$\begin{bmatrix} 2.5 & 6.5 \\ 4.2 & 3.2 \\ 2.0 & 1.5 \end{bmatrix} - \begin{bmatrix} 2.5 & 4.2 \\ 3.6 & 5.1 \\ 4.2 & 8.4 \end{bmatrix} = \begin{bmatrix} 0.0 & 2.3 \\ 0.6 & -1.9 \\ -2.2 & -6.9 \end{bmatrix}$$

(c) $$\begin{bmatrix} 2 & 0 & 4 \\ 1 & 1 & 2 \\ 3 & 2 & 4 \end{bmatrix} - \begin{bmatrix} 2 & 6 & 12 \\ 14 & 0 & 9 \\ 10 & 8 & 6 \end{bmatrix} = \begin{bmatrix} 0 & -6 & -8 \\ -13 & 1 & -7 \\ -7 & -6 & -2 \end{bmatrix}$$

(d) $$\begin{bmatrix} x \\ y \\ z \end{bmatrix} - \begin{bmatrix} x+1 \\ y^2 \\ 3z \end{bmatrix} = \begin{bmatrix} -1 \\ y - y^2 \\ -2z \end{bmatrix}$$

(e) Cannot be done since matrices $\boldsymbol{A}$ and $\boldsymbol{E}$ have different dimensions.

(f) $$\begin{bmatrix} 2a & 2b+c \\ c+d & d^2 \end{bmatrix} - \begin{bmatrix} a & b \\ c & d \end{bmatrix} = \begin{bmatrix} a & b+c \\ d & d^2 - d \end{bmatrix}$$

Self-Assessment Questions 7.2

1. Do you think that matrix addition commutes? That is to say is it always true that $\boldsymbol{A} + \boldsymbol{B} = \boldsymbol{B} + \boldsymbol{A}$ for any two $m \times n$ matrices $\boldsymbol{A}$ and $\boldsymbol{B}$? Justify your answer.
2. Do you think that matrix subtraction commutes? That is to say, is it always true that $\boldsymbol{A} - \boldsymbol{B} = \boldsymbol{B} - \boldsymbol{A}$ for any two $m \times n$ matrices $\boldsymbol{A}$ and $\boldsymbol{B}$? Justify your answer.
3. Are the operations of matrix addition and subtraction associative? That is to say, is it true that in general $(\boldsymbol{A} + \boldsymbol{B}) + \boldsymbol{C} = \boldsymbol{A} + (\boldsymbol{B} + \boldsymbol{C})$ and $(\boldsymbol{A} - \boldsymbol{B}) - \boldsymbol{C} = \boldsymbol{A} - (\boldsymbol{B} - \boldsymbol{C})$?

Exercise 7.2

1. Where possible calculate the following additions which refer to the matrices in Worked example 7.1 and Exercise 7.1.

   (a) $\boldsymbol{A} + \boldsymbol{F}$ (b) $\boldsymbol{L} + \boldsymbol{J}$ (c) $\boldsymbol{F} + \boldsymbol{E}$ (d) $\boldsymbol{J} + \boldsymbol{L}$ (e) $\boldsymbol{H} + \boldsymbol{M}$

2. With reference to the matrices in Exercise 7.1 and Worked example 7.1, where possible perform the following calculations.

   (a) $\boldsymbol{A} - \boldsymbol{F}$ (b) $\boldsymbol{G} - \boldsymbol{M}$ (c) $\boldsymbol{C} - \boldsymbol{J}$ (d) $\boldsymbol{G} - \boldsymbol{H}$ (e) $\boldsymbol{B} - \boldsymbol{M}$

7.3 Matrix multiplication

Matrix multiplication is quite a complicated calculation. We will approach it in stages. First we will consider the multiplication of a matrix by a scalar value, then multiplication that involves a simple row or column matrix, before considering the more general case.

Scalar multiplication

Referring back to the narrative from Section 7.1, let us suppose that the company sets a target that the sales in November should be double the sales achieved in June. The sales manager is asked to set particular targets for the sales of the different sized bolts in the three regions, based upon the sales figures for June. To do this he performs the following simple calculation:

Target for November = 2 × Sales in June

Using the matrix of sales values this becomes

$$\text{Target for November} = 2 \times \begin{array}{r} \\ \text{Small} \\ \text{Medium} \\ \text{Large} \end{array} \begin{array}{c} \begin{array}{ccc} \text{North} & \text{West} & \text{South} \end{array} \\ \begin{bmatrix} 100 & 200 & 70 \\ 60 & 60 & 90 \\ 120 & 90 & 110 \end{bmatrix} \end{array}$$

Dropping the labels for the columns and rows we get

$$\text{Target for November} = 2 \times \begin{bmatrix} 100 & 200 & 70 \\ 60 & 60 & 90 \\ 120 & 90 & 110 \end{bmatrix}$$

To perform this calculation Jim decides that each entry in the matrix for June must be multiplied by 2. Performing this calculation and replacing the labels on the rows and columns gives

	North	West	South
Small	200	400	140
Medium	120	120	180
Large	240	180	220

The previous calculation is an example of the multiplication of a matrix by a **scalar**. A scalar is some single value which may be an actual numeric value, as in the above case, or a variable representing a numeric value. In either case the process of scalar multiplication involves the multiplication of each element of the matrix by the scalar.

KEY POINT

In general for some $m \times n$ matrix A and some scalar value v we have

$$vA = v \begin{bmatrix} a_{11} & a_{12} & a_{13} & \cdots & a_{1n} \\ a_{21} & a_{22} & a_{23} & \cdots & a_{2n} \\ \vdots & \vdots & \vdots & & \vdots \\ a_{m1} & a_{m2} & a_{m3} & \cdots & a_{mn} \end{bmatrix}$$

giving

$$vA = \begin{bmatrix} va_{11} & va_{12} & va_{13} & \cdots & va_{1n} \\ va_{21} & va_{22} & va_{23} & \cdots & va_{2n} \\ \vdots & \vdots & \vdots & & \vdots \\ va_{m1} & va_{m2} & va_{m3} & \cdots & va_{mn} \end{bmatrix}$$

Worked examples

7.3 The following matrix shows the values of sales achieved by a company's representatives in three given regions. If the representatives are required to increase the value of their sales by 5% in each region, calculate the sales targets for each representative.

Sales achieved	North	West	South
Anne	200	220	260
Brian	160	140	240
Carol	120	100	180

Solution If the performance of the representatives is to increase by 5% then the sales targets can be calculated by multiplying current sales values by 1.05. This can be achieved by multiplying the given matrix of sales values by the scalar 1.05.

$$\text{Sales targets} = 1.05 \times \begin{bmatrix} 200 & 220 & 260 \\ 160 & 140 & 240 \\ 120 & 100 & 180 \end{bmatrix}$$

$$= \begin{bmatrix} 210 & 231 & 273 \\ 168 & 147 & 252 \\ 126 & 105 & 189 \end{bmatrix}$$

Hence the new sales targets are

$$\begin{array}{l} \\ \text{Anne} \\ \text{Brian} \\ \text{Carol} \end{array} \begin{array}{c} \text{North} \quad \text{West} \quad \text{South} \\ \begin{bmatrix} 210 & 231 & 273 \\ 168 & 147 & 252 \\ 126 & 105 & 189 \end{bmatrix} \end{array}$$

7.4 With reference to the matrices in Worked example 7.1 and Exercise 7.1 perform the following calculations.

(a) $3\boldsymbol{J} + \boldsymbol{L}$ (b) $\boldsymbol{A} - 2\boldsymbol{F}$

Solution (a)

$$3\boldsymbol{J} + \boldsymbol{L} = 3 \begin{bmatrix} x \\ y \\ z \end{bmatrix} + \begin{bmatrix} x+1 \\ y^2 \\ 3z \end{bmatrix} = \begin{bmatrix} 3x \\ 3y \\ 3z \end{bmatrix} + \begin{bmatrix} x+1 \\ y^2 \\ 3z \end{bmatrix}$$

$$3\boldsymbol{J} + \boldsymbol{L} = \begin{bmatrix} 4x+1 \\ 3y+y^2 \\ 6z \end{bmatrix}$$

(b)

$$\boldsymbol{A} - 2\boldsymbol{F} = \begin{bmatrix} 2 & 6 & 12 \\ 14 & 0 & 9 \\ 10 & 8 & 6 \end{bmatrix} - 2 \begin{bmatrix} 2 & 0 & 4 \\ 1 & 1 & 2 \\ 3 & 2 & 4 \end{bmatrix}$$

$$= \begin{bmatrix} 2 & 6 & 12 \\ 14 & 0 & 9 \\ 10 & 8 & 6 \end{bmatrix} - \begin{bmatrix} 4 & 0 & 8 \\ 2 & 2 & 4 \\ 6 & 4 & 8 \end{bmatrix}$$

$$\boldsymbol{A} - 2\boldsymbol{F} = \begin{bmatrix} -2 & 6 & 4 \\ 12 & -2 & 5 \\ 4 & 4 & -2 \end{bmatrix}$$

The previous examples illustrate the fact that the multiplication of a matrix by a scalar has a higher priority than either matrix addition or subtraction. In other words, when evaluating an expression you should carry out scalar multiplication before adding or subtracting matrices.

Worked example

7.5 For the matrices

$$\boldsymbol{A} = \begin{bmatrix} 3 & 4 \\ 2 & 7 \end{bmatrix} \quad \boldsymbol{B} = \begin{bmatrix} 2 & 5 \\ 4 & 6 \end{bmatrix} \quad \boldsymbol{C} = \begin{bmatrix} 15 & 27 \\ 18 & 39 \end{bmatrix}$$

show that the following relationship holds: $3(\boldsymbol{A} + \boldsymbol{B}) = \boldsymbol{C}$

Solution In this example the brackets ensure that the operation of addition has priority over the multiplication by a scalar. Now

$$\boldsymbol{A} + \boldsymbol{B} = \begin{bmatrix} 3 & 4 \\ 2 & 7 \end{bmatrix} + \begin{bmatrix} 2 & 5 \\ 4 & 6 \end{bmatrix} = \begin{bmatrix} 5 & 9 \\ 6 & 13 \end{bmatrix}$$

So

$$3(\boldsymbol{A} + \boldsymbol{B}) = 3 \begin{bmatrix} 5 & 9 \\ 6 & 13 \end{bmatrix} = \begin{bmatrix} 15 & 27 \\ 18 & 39 \end{bmatrix}$$

Now

$$\boldsymbol{C} = \begin{bmatrix} 15 & 27 \\ 18 & 39 \end{bmatrix}$$

Since the values in corresponding positions in the two matrices above are equal the matrices themselves are equal. Hence we can write $3(\boldsymbol{A} + \boldsymbol{B}) = \boldsymbol{C}$.

Multiplication involving a row matrix

Jim the sales manager, who we met in Section 7.1, has been requested to provide details of the value of the sales of bolts in the three regions of the country for the month of June. Jim has data on the numbers of bolts sold in June which he has recorded in a matrix as shown below.

Sales in June	North	West	South
Small	100	200	70
Medium	60	60	90
Large	120	90	110

Jim also has data on the unit costs of the three sizes of bolt, again stored in matrix form, but this time as a simple row matrix as shown below.

Unit costs of bolts (in pence)

$$\begin{matrix} \text{Small} & \text{Medium} & \text{Large} \end{matrix}$$
$$\begin{bmatrix} 50 & 60 & 70 \end{bmatrix}$$

In order to calculate the value of the sales Jim knows that he must multiply unit costs by quantity sold. He therefore performs the following calculation.

unit costs × sales for June

Writing this using matrices we get

$$\begin{matrix} \text{Small} & \text{Medium} & \text{Large} \\ \end{matrix} \qquad \begin{matrix} & \text{North} & \text{West} & \text{South} \end{matrix}$$
$$\begin{bmatrix} 50 & 60 & 70 \end{bmatrix} \times \begin{matrix} \text{Small} \\ \text{Medium} \\ \text{Large} \end{matrix} \begin{bmatrix} 100 & 200 & 70 \\ 60 & 60 & 90 \\ 120 & 90 & 110 \end{bmatrix}$$

In performing this calculation the sales quantities for each size of bolt must be multiplied by the appropriate unit cost and the results for the three sizes added together to give the value of sales in a particular region. The results of this calculation are as follows.

$$\begin{matrix} \text{North} & \text{West} \\ [50 \times 100 + 60 \times 60 + 70 \times 120 & 50 \times 200 + 60 \times 60 + 70 \times 90 \end{matrix}$$
$$\begin{matrix} \text{South} \\ 50 \times 70 + 60 \times 90 + 70 \times 110] \end{matrix}$$

Performing the arithmetic we get

$$\begin{matrix} \text{North} & \text{West} & \text{South} \end{matrix}$$
$$\begin{bmatrix} 17000 & 19900 & 16600 \end{bmatrix}$$

Let us look at this calculation a little more closely. The quantities of sales of the different sized bolts, in the Northern region, appear in the first column of the 'Sales for June' matrix. That is,

Sales in June

$$\begin{matrix} & \text{North} & \text{West} & \text{South} \end{matrix}$$
$$\begin{matrix} \text{Small} \\ \text{Medium} \\ \text{Large} \end{matrix} \begin{bmatrix} 100 & & \\ 60 & & \\ 120 & & \end{bmatrix}$$

The unit costs for the three sizes of bolt are held in the row matrix

Unit costs

$$\begin{matrix} \text{Small} & \text{Medium} & \text{Large} \end{matrix}$$
$$\begin{bmatrix} 50 & 60 & 70 \end{bmatrix}$$

So to calculate the value of the sales of small bolts in the North we multiply the first entry in the column by the first entry in the row matrix. In like manner, the value for the sales of medium bolts is calculated by multiplying the second entries together and for large bolts by multiplying the third entries together.

The result is

Sales values for bolts

for small bolts	50×100
for medium bolts	60×60
for large bolts	70×120

The sum of these three calculations gives the total value of all sales in the North. That is,

Total sales in North

$$(50 \times 100) + (60 \times 60) + (70 \times 120)$$

To calculate the total sales for the West and for the South the above calculation is repeated but this time using the second and third columns of the sales matrix respectively. If we were to remove the labels the above calculation would look like

$$\begin{bmatrix} 50 & 60 & 70 \end{bmatrix} \begin{bmatrix} 100 & 200 & 70 \\ 60 & 60 & 90 \\ 120 & 90 & 110 \end{bmatrix}$$

giving

$$[50 \times 100 + 60 \times 60 + 70 \times 120 \quad 50 \times 200 + 60 \times 60 + 70 \times 90 \quad 50 \times 70 + 60 \times 90 + 70 \times 110]$$

Performing the arithmetic we get

$$\begin{bmatrix} 17000 & 19900 & 16600 \end{bmatrix}$$

What we have here is an example of the multiplication of a row matrix by a 3×3 matrix. An important thing to note is that the row matrix has the same number of **columns** as there are **rows** in the 3×3 matrix. This condition must hold if multiplication of the matrices is to be possible.

KEY POINT

In general if two matrices are to be multiplied together, the number of columns in the left-hand matrix must be the same as the number of rows in the right-hand matrix.

The result of such a multiplication will be a matrix which has the same number of rows as the left-hand matrix and the same number of columns as the right-hand matrix.

Worked example

7.6 Using the matrices

$$A = \begin{bmatrix} 2 & 3 \\ 3 & 2 \\ 4 & 6 \end{bmatrix} \quad B = \begin{bmatrix} 2 \\ 3 \\ 7 \end{bmatrix} \quad C = \begin{bmatrix} x+1 & y \\ x-2 & 2y \\ x & y-4 \end{bmatrix} \quad D = \begin{bmatrix} a & c \\ b & a \end{bmatrix}$$

$$E = [3 \quad 6 \quad 4] \quad F = [4 \quad 5] \quad G = [x \quad y \quad z]$$

where possible perform the following calculations.

(a) $\boldsymbol{EA}$ (b) $\boldsymbol{BC}$ (c) $\boldsymbol{EB}$ (d) $\boldsymbol{AG}$ (e) $\boldsymbol{FD}$

Solution (a)

$$\boldsymbol{EA} = [3 \quad 6 \quad 4] \begin{bmatrix} 2 & 3 \\ 3 & 2 \\ 4 & 6 \end{bmatrix}$$

$$= [(3 \times 2) + (6 \times 3) + (4 \times 4) \quad (3 \times 3) + (6 \times 2) + (4 \times 6)]$$
$$= [40 \quad 45]$$

This is an example of a matrix $\boldsymbol{E}$ consisting of a single row being **postmultiplied**, i.e. multiplied on the right, by a 3×2 matrix.

(b) $$\boldsymbol{BC} = \begin{bmatrix} 2 \\ 3 \\ 7 \end{bmatrix} \begin{bmatrix} x+1 & y \\ x-2 & 2y \\ x & y-4 \end{bmatrix}$$

Cannot be done since $\boldsymbol{B}$ does not have the same number of columns as $\boldsymbol{C}$ has rows.

(c) $$\boldsymbol{EB} = [3 \quad 6 \quad 4] \begin{bmatrix} 2 \\ 3 \\ 7 \end{bmatrix}$$

$$= [(3 \times 2) + (6 \times 3) + (4 \times 7)]$$
$$= [52]$$

(An example of a 1×1 matrix)

(d) $$\boldsymbol{AG} = \begin{bmatrix} 2 & 3 \\ 3 & 2 \\ 4 & 6 \end{bmatrix} [x \quad y \quad z]$$

Cannot be done since $\boldsymbol{A}$ has two columns while $\boldsymbol{G}$ has only one row.

(e) $$\boldsymbol{FD} = [4 \quad 5] \begin{bmatrix} a & c \\ b & a \end{bmatrix} = [4a + 5b \quad 4c + 5a]$$

Multiplication involving a column matrix

Our fictitious engineering company pays its sales representatives a monthly commission based upon the value of sales they achieve. The rate at which the commission is paid varies from region to region. The commission on sales in the North is 2% , in the West it is 3% while in the South it is just 1%. If the following matrix holds the values of sales, expressed in pounds, achieved by the company's representatives for the month of May,

Sales for May in £s

$$\begin{array}{l} \\ \text{Anne} \\ \text{Brian} \\ \text{Carol} \end{array} \begin{array}{c} \text{North West South} \\ \begin{bmatrix} 200 & 220 & 260 \\ 160 & 140 & 240 \\ 120 & 100 & 180 \end{bmatrix} \end{array}$$

how can we calculate the commissions owing to the representatives? What we need to do for a particular representative is to multiply the value of the sales achieved in each of the regions by the appropriate commission rate and then to add the three values together. If we record the commission rates in the simple column matrix

$$\begin{array}{c} \text{North} \\ \text{West} \\ \text{South} \end{array} \begin{bmatrix} 2\% \\ 3\% \\ 1\% \end{bmatrix}$$

then the calculation can be represented as

$$\begin{array}{l} \\ \text{Anne} \\ \text{Brian} \\ \text{Carol} \end{array} \begin{array}{c} \text{North West South} \\ \begin{bmatrix} 200 & 220 & 260 \\ 160 & 140 & 240 \\ 120 & 100 & 180 \end{bmatrix} \end{array} \times \begin{array}{c} \text{North} \\ \text{West} \\ \text{South} \end{array} \begin{bmatrix} 2\% \\ 3\% \\ 1\% \end{bmatrix}$$

Writing the percentages as decimal values and removing some of the labelling we get

$$\begin{array}{l} \\ \text{Anne} \\ \text{Brian} \\ \text{Carol} \end{array} \begin{array}{c} \text{North West South} \\ \begin{bmatrix} 200 & 220 & 260 \\ 160 & 140 & 240 \\ 120 & 100 & 180 \end{bmatrix} \end{array} \begin{array}{c} \\ \begin{bmatrix} 0.02 \\ 0.03 \\ 0.01 \end{bmatrix} \end{array}$$

To carry out this calculation we multiply each element from a given row of the left-hand matrix by the corresponding element in the column matrix. The three results are then added together. This calculation is repeated for each row of the left-hand matrix. If we just consider the calculation relating to Anne we get the following:

$$\begin{array}{l} \\ \text{Anne} \\ \text{Brian} \\ \text{Carol} \end{array} \begin{array}{c} \text{North West South} \\ \begin{bmatrix} 200 & 220 & 260 \\ & & \\ & & \end{bmatrix} \end{array} \begin{array}{c} \\ \begin{bmatrix} 0.02 \\ 0.03 \\ 0.01 \end{bmatrix} \end{array}$$

which gives

$$\begin{array}{l} \text{Anne} \\ \text{Brian} \\ \text{Carol} \end{array} \begin{bmatrix} (200 \times 0.02) + (220 \times 0.03) + (260 \times 0.01) \\ \\ \\ \end{bmatrix}$$

If we now include the values for the other two representatives we get

$$\begin{array}{l} \\ \text{Anne} \\ \text{Brian} \\ \text{Carol} \end{array} \begin{array}{c} \begin{array}{ccc} \text{North} & \text{West} & \text{South} \end{array} \\ \begin{bmatrix} 200 & 220 & 260 \\ 160 & 140 & 240 \\ 120 & 100 & 180 \end{bmatrix} \end{array} \begin{array}{c} \\ \begin{bmatrix} 0.02 \\ 0.03 \\ 0.01 \end{bmatrix} \end{array}$$

which gives

$$\begin{array}{l} \text{Anne} \\ \text{Brian} \\ \text{Carol} \end{array} \begin{bmatrix} (200 \times 0.02) + (220 \times 0.03) + (260 \times 0.01) \\ (160 \times 0.02) + (140 \times 0.03) + (240 \times 0.01) \\ (120 \times 0.02) + (100 \times 0.03) + (180 \times 0.01) \end{bmatrix}$$

Performing the multiplications gives

$$\begin{array}{l} \text{Anne} \\ \text{Brian} \\ \text{Carol} \end{array} \begin{bmatrix} 4.0 + 6.6 + 2.6 \\ 3.2 + 4.2 + 2.4 \\ 2.4 + 3.0 + 1.8 \end{bmatrix}$$

Finally we get

$$\begin{array}{l} \text{Anne} \\ \text{Brian} \\ \text{Carol} \end{array} \begin{bmatrix} 13.2 \\ 9.8 \\ 7.2 \end{bmatrix}$$

So the commission received by the representatives in May was as follows: Anne £13.20, Brian £9.80 and Carol £7.20. If we remove all of the labelling from the previous calculation we get

$$\begin{bmatrix} 200 & 220 & 260 \\ 160 & 140 & 240 \\ 120 & 100 & 180 \end{bmatrix} \begin{bmatrix} 0.02 \\ 0.03 \\ 0.01 \end{bmatrix}$$

Performing the matrix multiplication we get

$$\begin{bmatrix} (200 \times 0.02) + (220 \times 0.03) + (260 \times 0.01) \\ (160 \times 0.02) + (140 \times 0.03) + (240 \times 0.01) \\ (120 \times 0.02) + (100 \times 0.03) + (180 \times 0.01) \end{bmatrix}$$

Performing the arithmetic gives

$$\begin{bmatrix} 13.2 \\ 9.8 \\ 7.2 \end{bmatrix}$$

This is an example of the multiplication of a 3×3 matrix by a column matrix. Notice that in order to be able to perform this calculation the left-hand matrix has the same number of columns as the right-hand matrix has rows. Note also that the resultant matrix has the same number of rows as the left-hand matrix and the same number of columns as the right-hand matrix.

Here are some more examples of multiplications involving a column matrix.

Worked example

7.7 Using the matrices

$$A = \begin{bmatrix} 2 & 3 & 4 \\ 5 & 6 & 7 \end{bmatrix} \quad B = \begin{bmatrix} 2 & 5 & 4 \\ 4 & 6 & 2 \\ 3 & 2 & 1 \end{bmatrix} \quad C = \begin{bmatrix} x & y \\ 2x & -y \end{bmatrix}$$

$$D = [a \quad b \quad c]$$

$$E = \begin{bmatrix} 2 \\ 3 \\ 4 \end{bmatrix} \quad F = \begin{bmatrix} 2 \\ 1 \end{bmatrix} \quad G = \begin{bmatrix} x \\ y \\ z \end{bmatrix} \quad H = \begin{bmatrix} a \\ b \end{bmatrix}$$

where possible perform the following calculations.

(a) AE (b) CE (c) DE (e) CH

Solution

(a) $$AE = \begin{bmatrix} 2 & 3 & 4 \\ 5 & 6 & 7 \end{bmatrix} \begin{bmatrix} 2 \\ 3 \\ 4 \end{bmatrix} = \begin{bmatrix} 2 \times 2 + 3 \times 3 + 4 \times 4 \\ 5 \times 2 + 6 \times 3 + 7 \times 4 \end{bmatrix} = \begin{bmatrix} 29 \\ 56 \end{bmatrix}$$

This is an example of a matrix E, consisting of a single column, being **premultiplied**, i.e. multiplied on the left, by a 2×3 matrix.

(b) CE cannot be calculated since C has 2 columns while E has 3 rows.

(c) $$DE = [a \quad b \quad c] \begin{bmatrix} 2 \\ 3 \\ 4 \end{bmatrix} = [2a + 3b + 4c]$$

(d) $$CH = \begin{bmatrix} x & y \\ 2x & -y \end{bmatrix} \begin{bmatrix} a \\ b \end{bmatrix} = \begin{bmatrix} ax + by \\ 2ax - by \end{bmatrix}$$

A more general case

In this section we will consider a more general example of matrix multiplication. As a motivation let us return to the small engineering company and consider the problem of calculating the costs of overheads incurred in selling bolts in the three regions of the country. For this example we will regard the costs of deliveries and warehousing as the two main overheads. The sales manager records these costs, which vary from region to region, in the following matrix:

Costs per unit in pence

$$\begin{array}{l} \\ \text{North} \\ \text{West} \\ \text{South} \end{array} \begin{array}{c} \begin{array}{cc} \text{Delivery} & \text{Warehousing} \end{array} \\ \begin{bmatrix} 0.3 & 0.01 \\ 0.2 & 0.02 \\ 0.2 & 0.04 \end{bmatrix} \end{array}$$

From this matrix we see that the cost of delivering a single bolt in the Northern region is assumed to be 0.3p, while the warehousing cost for a single bolt in the South is 0.04p. If requested for the total cost of the overheads associated with the sales of bolts in the month of June the sales manager first retrieves the matrix giving the breakdown of sales totals for that month. This is shown below.

Sales in June

$$\begin{array}{l} \\ \text{Small} \\ \text{Medium} \\ \text{Large} \end{array} \begin{array}{c} \begin{array}{ccc} \text{North} & \text{West} & \text{South} \end{array} \\ \begin{bmatrix} 100 & 200 & 70 \\ 60 & 60 & 90 \\ 120 & 90 & 110 \end{bmatrix} \end{array}$$

He then reasons in the following way. To calculate the overhead costs incurred in June, the costs per unit must be multiplied by the number of units sold. This would suggest that the two matrices given above should be multiplied together. Now from the previous two sections we know that if two matrices are to be multiplied then the number of columns in the left-hand matrix must be the same as the number of rows in the right-hand matrix. We see that the matrix recording unit costs has three rows while the matrix recording sales has three columns. This fact tells us that the following calculation is possible.

Sales Costs

$$\begin{array}{l} \\ \text{Small} \\ \text{Medium} \\ \text{Large} \end{array} \begin{array}{c} \begin{array}{ccc} \text{North} & \text{West} & \text{South} \end{array} \\ \begin{bmatrix} 100 & 200 & 70 \\ 60 & 60 & 90 \\ 120 & 90 & 110 \end{bmatrix} \end{array} \times \begin{array}{l} \\ \text{North} \\ \text{West} \\ \text{South} \end{array} \begin{array}{c} \begin{array}{cc} \text{Delivery} & \text{Warehousing} \end{array} \\ \begin{bmatrix} 0.3 & 0.01 \\ 0.2 & 0.02 \\ 0.2 & 0.04 \end{bmatrix} \end{array}$$

To simplify things a little let us consider how to perform this calculation for just small bolts. We have

Sales Costs

$$\begin{array}{l} \\ \text{Small} \\ \text{Medium} \\ \text{Large} \end{array} \begin{array}{c} \begin{array}{ccc} \text{North} & \text{West} & \text{South} \end{array} \\ \begin{bmatrix} 100 & 200 & 70 \\ & & \\ & & \end{bmatrix} \end{array} \times \begin{array}{l} \\ \text{North} \\ \text{West} \\ \text{South} \end{array} \begin{array}{c} \begin{array}{cc} \text{Delivery} & \text{Warehousing} \end{array} \\ \begin{bmatrix} 0.3 & 0.01 \\ 0.2 & 0.02 \\ 0.2 & 0.04 \end{bmatrix} \end{array}$$

Multiplying the entries in the row relating to small bolts with the corresponding entries in the column relating to deliveries and adding the results together we get

Costs of overheads

$$\begin{array}{r} \\ \text{Small} \\ \text{Medium} \\ \text{Large} \end{array} \begin{array}{c} \text{Deliveries} \qquad\qquad \text{Warehousing} \\ \left[\begin{array}{cc} 100 \times 0.3 + 200 \times 0.2 + 70 \times 0.2 & \qquad\qquad \\ & \\ & \end{array}\right] \end{array}$$

If we now include the calculation relating to the warehousing costs we get

Costs of overheads

$$\begin{array}{r} \\ \text{Small} \\ \text{Medium} \\ \text{Large} \end{array} \begin{array}{c} \text{Deliveries} \qquad\qquad\qquad \text{Warehousing} \\ \left[\begin{array}{cc} 100 \times 0.3 + 200 \times 0.2 + 70 \times 0.2 & 100 \times 0.01 + 200 \times 0.02 + 70 \times 0.04 \\ & \\ & \end{array}\right] \end{array}$$

Including the calculations for the other two sizes of bolt gives

Costs of overheads

$$\begin{array}{r} \\ \text{Small} \\ \text{Medium} \\ \text{Large} \end{array} \begin{array}{c} \text{Deliveries} \qquad\qquad\qquad \text{Warehousing} \\ \left[\begin{array}{cc} 100 \times 0.3 + 200 \times 0.2 + 70 \times 0.2 & 100 \times 0.01 + 200 \times 0.02 + 70 \times 0.04 \\ 60 \times 0.3 + 60 \times 0.2 + 90 \times 0.2 & 60 \times 0.01 + 60 \times 0.02 + 90 \times 0.04 \\ 120 \times 0.3 + 90 \times 0.2 + 110 \times 0.2 & 120 \times 0.01 + 90 \times 0.02 + 110 \times 0.04 \end{array}\right] \end{array}$$

Doing the arithmetic gives

Costs of overheads (in pence)

$$\begin{array}{r} \\ \text{Small} \\ \text{Medium} \\ \text{Large} \end{array} \begin{array}{c} \text{Deliveries} \quad \text{Warehousing} \\ \left[\begin{array}{cc} 84 & 7.8 \\ 48 & 5.4 \\ 76 & 7.4 \end{array}\right] \end{array}$$

We have ended up with a matrix which shows the overhead costs for deliveries and warehousing for the three sizes of bolt. If we remove the labelling from the above calculations we have

$$\begin{bmatrix} 100 & 200 & 70 \\ 60 & 60 & 90 \\ 120 & 90 & 110 \end{bmatrix} \begin{bmatrix} 0.3 & 0.01 \\ 0.2 & 0.02 \\ 0.2 & 0.04 \end{bmatrix}$$

$$= \begin{bmatrix} 100 \times 0.3 + 200 \times 0.2 + 70 \times 0.2 & 100 \times 0.01 + 200 \times 0.02 + 70 \times 0.04 \\ 60 \times 0.3 + 60 \times 0.2 + 90 \times 0.2 & 60 \times 0.01 + 60 \times 0.02 + 90 \times 0.04 \\ 120 \times 0.3 + 90 \times 0.2 + 110 \times 0.2 & 120 \times 0.01 + 90 \times 0.02 + 110 \times 0.04 \end{bmatrix}$$

Performing the arithmetic gives

$$\begin{bmatrix} 84 & 7.8 \\ 48 & 5.4 \\ 76 & 7.4 \end{bmatrix}$$

This is a more general example of matrix multiplication than those we have seen before. Instead of the multiplication involving a matrix with a single column or a single row, both of the matrices have more than one row and column. However, it is still the case that the number of columns in the left-hand matrix is the same as the number of rows in the right-hand matrix.

KEY POINT

In general, if we wish to perform the calculation $\boldsymbol{AB}$, where $\boldsymbol{A}$ is an $m \times n$ matrix, i.e. it has m rows and n columns, then $\boldsymbol{B}$ must have n rows and one or more columns. Let us assume that $\boldsymbol{B}$ has k columns, i.e. it is an $n \times k$ matrix, then we have

$$\boldsymbol{AB} = \begin{bmatrix} a_{11} & a_{12} & a_{13} & \cdots & a_{1n} \\ a_{21} & a_{22} & a_{23} & \cdots & a_{2n} \\ \vdots & \vdots & \vdots & & \vdots \\ a_{m1} & a_{m2} & a_{m3} & \cdots & a_{mn} \end{bmatrix} \begin{bmatrix} b_{11} & b_{12} & b_{13} & \cdots & b_{1k} \\ b_{21} & b_{22} & b_{23} & \cdots & b_{2k} \\ \vdots & \vdots & \vdots & & \vdots \\ b_{n1} & b_{n2} & b_{n3} & \cdots & b_{nk} \end{bmatrix}$$

If the result of this calculation is a matrix $\boldsymbol{C}$ then we know that $\boldsymbol{C}$ must have m rows and k columns. So we have

$$\boldsymbol{C} = \begin{bmatrix} c_{11} & c_{12} & c_{13} & \cdots & c_{1k} \\ c_{21} & c_{22} & c_{23} & \cdots & c_{2k} \\ \vdots & \vdots & \vdots & & \vdots \\ c_{m1} & c_{m2} & c_{m3} & \cdots & c_{mk} \end{bmatrix}$$

The value of some arbitrary element c_{ij} of this result matrix is obtained by multiplying the elements of the ith row of matrix $\boldsymbol{A}$ with the corresponding elements of the jth column of matrix $\boldsymbol{B}$ and adding the resulting values together. That is,

$$c_{ij} = a_{i1}b_{1j} + a_{i2}b_{2j} + \cdots + a_{in}b_{nj}$$

Worked example

7.8 With reference to the matrices

$$\boldsymbol{A} = \begin{bmatrix} a & b \\ c & d \end{bmatrix} \quad \boldsymbol{B} = \begin{bmatrix} 2 & 3 \\ 4 & 5 \end{bmatrix} \quad \boldsymbol{C} = \begin{bmatrix} 1 & 2 & 3 \\ 4 & 5 & 6 \end{bmatrix}$$

perform the following calculations if possible.

(a) $\boldsymbol{BA}$ (b) $\boldsymbol{A}^2$ (i.e. $\boldsymbol{A} \times \boldsymbol{A}$) (c) $\boldsymbol{CB}$ (d) $\boldsymbol{BC}$ (e) $\boldsymbol{AB}$

Solution (a) $BA = \begin{bmatrix} 2 & 3 \\ 4 & 5 \end{bmatrix} \begin{bmatrix} a & b \\ c & d \end{bmatrix}$ which gives $BA = \begin{bmatrix} 2a+3c & 2b+3d \\ 4a+5c & 4b+5d \end{bmatrix}$

(b) $A^2 = \begin{bmatrix} a & b \\ c & d \end{bmatrix} \begin{bmatrix} a & b \\ c & d \end{bmatrix}$ which gives $A^2 = \begin{bmatrix} a^2+bc & ab+bd \\ ca+dc & cb+d^2 \end{bmatrix}$

(c) It is not possible to perform this calculation since C has 3 columns while B has only 2 rows.

(d) $BC = \begin{bmatrix} 2 & 3 \\ 4 & 5 \end{bmatrix} \begin{bmatrix} 1 & 2 & 3 \\ 4 & 5 & 6 \end{bmatrix} = \begin{bmatrix} 2+12 & 4+15 & 6+18 \\ 4+20 & 8+25 & 12+30 \end{bmatrix}$

which gives $BC = \begin{bmatrix} 14 & 19 & 24 \\ 24 & 33 & 42 \end{bmatrix}$

(e) $AB = \begin{bmatrix} a & b \\ c & d \end{bmatrix} \begin{bmatrix} 2 & 3 \\ 4 & 5 \end{bmatrix} = \begin{bmatrix} 2a+4b & 3a+5b \\ 2c+4d & 3c+5d \end{bmatrix}$

From Worked examples 7.8 (a) and 7.8 (e) we see that $AB \neq BA$ and so we may conclude that in general matrix multiplication is not commutative, that is to say for any two matrices A and B it is not generally true that $AB = BA$.

From Worked examples 7.8 (c) and 7.8 (d) we see that though it may be possible to calculate the product BC, for two given matrices B and C, it does not necessarily follow that it is possible to calculate the product CB.

Self-Assessment Questions 7.3

1. Do you think it is generally true that $n(A + B) = nA + nB$, where n is a scalar value? Justify your answer.
2. Do you think that it is generally true that $AB + AC = A(B + C)$? Justify your answer.

Exercise 7.3

1. The costs of running three departments of a particular organization are shown in the matrix below. The data values represent thousands of pounds.

	DeptA	DeptB	DeptC
Materials	100	80	20
Wages	90	60	120
Overheads	40	50	40

If all departments must reduce expenditure in all areas by 10%, calculate the matrix which shows target costs.

2. For the matrices

$$A = \begin{bmatrix} 2 & 3 \\ 6 & 5 \end{bmatrix} \quad B = \begin{bmatrix} 2 & 4 \\ 5 & 3 \end{bmatrix} \quad C = \begin{bmatrix} 10 & 17 \\ 28 & 21 \end{bmatrix}$$

show that the following relationship holds:

$$3\boldsymbol{A} + 2\boldsymbol{B} = \boldsymbol{C}$$

3. Find the values of the variables x, y and z in the following equation.

$$\begin{bmatrix} 10 & 8 \\ 8 & 14 \end{bmatrix} = \begin{bmatrix} 5 & x \\ 2 & z \end{bmatrix} + 2\begin{bmatrix} y & 3 \\ 3 & 6 \end{bmatrix}$$

4. With reference to the matrices in Worked example 7.6 where possible perform the following calculations.

(a) $\boldsymbol{EC}$ (b) $\boldsymbol{AD}$ (c) $\boldsymbol{GA}$ (d) $\boldsymbol{GB}$ (e) $\boldsymbol{BG}$

5. With reference to the matrices in Worked example 7.7 where possible perform the following calculations.

(a) $\boldsymbol{BG}$ (b) $\boldsymbol{BE}$ (c) $\boldsymbol{FH}$ (d) $\boldsymbol{DG}$ (e) $\boldsymbol{CF}$

6. With reference to the matrices

$$D = \begin{bmatrix} 6 & 5 \\ 4 & 7 \\ 2 & 3 \end{bmatrix} \quad E = \begin{bmatrix} x & y \\ 2x & 3y \end{bmatrix} \quad F = \begin{bmatrix} 2 & 5 & 3 \\ 3 & 4 & 2 \\ 4 & 2 & 1 \end{bmatrix}$$

and those matrices in Worked example 7.8 perform the following calculations if possible.

(a) $\boldsymbol{DA}$ (b) $\boldsymbol{B}^2$ (c) $\boldsymbol{DF}$ (d) $\boldsymbol{FD}$ (e) $\boldsymbol{BE}$ (f) $\boldsymbol{C}^2$

7.4 The identity matrix

From the algebra of real numbers you will be familiar with equations of the form

$$2x + 6 = 9$$

To solve such an equation we must find a value for the variable x which makes the expression true. Similarly we may create equations that involve matrices. For example, consider the problem of solving the equation

$$\begin{bmatrix} a & b \\ c & d \end{bmatrix} \begin{bmatrix} 2 & 3 \\ 4 & 5 \end{bmatrix} = \begin{bmatrix} 2 & 3 \\ 4 & 5 \end{bmatrix}$$

To find a solution to this equation we need to identify values for the variables a, b, c and d which will make the expression true.

If we perform the multiplication on the left-hand side of the equation we get

$$\begin{bmatrix} 2a+4b & 3a+5b \\ 2c+4d & 3c+5d \end{bmatrix} = \begin{bmatrix} 2 & 3 \\ 4 & 5 \end{bmatrix}$$

If this equality is to hold then the corresponding entries in the two matrices must be equal. Hence the following equations must hold:

(i) $2a + 4b = 2$ (ii) $3a + 5b = 3$
(iii) $2c + 4d = 4$ (iv) $3c + 5d = 5$

Equations (i) and (ii) will hold if $a = 1$ and $b = 0$, while equations (iii) and (iv) will hold if $c = 0$ and $d = 1$. This suggests that

$$\begin{bmatrix} 1 & 0 \\ 0 & 1 \end{bmatrix} \begin{bmatrix} 2 & 3 \\ 4 & 5 \end{bmatrix} = \begin{bmatrix} 2 & 3 \\ 4 & 5 \end{bmatrix}$$

Performing the multiplication on the left-hand side of the equation we get

$$\begin{bmatrix} 2 & 3 \\ 4 & 5 \end{bmatrix} = \begin{bmatrix} 2 & 3 \\ 4 & 5 \end{bmatrix}$$

So the equality holds when the unknown matrix is equal to

$$\begin{bmatrix} 1 & 0 \\ 0 & 1 \end{bmatrix}$$

This matrix is a particular example of an **identity matrix**. Any 2×2 matrix multiplied by this identity matrix will remain unchanged.

If we consider some general 3×3 matrix, say

$$\begin{bmatrix} a & b & c \\ d & e & f \\ g & h & i \end{bmatrix}$$

then the following equation holds:

$$\begin{bmatrix} 1 & 0 & 0 \\ 0 & 1 & 0 \\ 0 & 0 & 1 \end{bmatrix} \begin{bmatrix} a & b & c \\ d & e & f \\ g & h & i \end{bmatrix} = \begin{bmatrix} a & b & c \\ d & e & f \\ g & h & i \end{bmatrix}$$

We see that the 3×3 identity matrix is

$$\begin{bmatrix} 1 & 0 & 0 \\ 0 & 1 & 0 \\ 0 & 0 & 1 \end{bmatrix}$$

Similarly the 4×4 identity matrix is

$$\begin{bmatrix} 1 & 0 & 0 & 0 \\ 0 & 1 & 0 & 0 \\ 0 & 0 & 1 & 0 \\ 0 & 0 & 0 & 1 \end{bmatrix}$$

KEY POINT

In general any **square** matrix which contains only 1s in the locations along the diagonal running from the top left corner of the matrix to the bottom right and 0s in all other locations is an identity matrix.

We represent the general identity matrix by the letter $\boldsymbol{I}$. For any square matrix $\boldsymbol{A}$ we can write

$$\boldsymbol{IA} = \boldsymbol{AI} = \boldsymbol{A}$$

Self-Assessment Question 7.4

Can you explain why the matrix $\boldsymbol{A}$ must be a square matrix for the equation $\boldsymbol{IA} = \boldsymbol{AI} = \boldsymbol{A}$ to hold?

7.5 Elementary row operations

In Section 7.7 we shall see that the ability to modify the values held in specific rows of a matrix can be used in solving systems of linear equations. In particular the following operations prove useful:

- Multiplying a row of a matrix by a scalar value.
- Adding one row of a matrix to another row.
- Exchanging two rows of a matrix.

These operations are known as **elementary row operations** and each may be achieved by **premultiplying** the given matrix by an **elementary matrix**. An elementary matrix is generated by performing an elementary row operation on an identity matrix. For example, the elementary matrix

$$\begin{bmatrix} 1 & 0 & 0 \\ 0 & 5 & 0 \\ 0 & 0 & 1 \end{bmatrix}$$

has been generated by multiplying row 2 of the 3×3 identity matrix by 5.

We shall now consider each of the elementary row operations in turn.

Multiplying a row by a scalar

The multiplication of row 2 of the matrix

$$\begin{bmatrix} a & b & c \\ d & e & f \\ g & h & i \end{bmatrix}$$

by the scalar 4 can be achieved by premultiplying the matrix by the elementary matrix

$$\begin{bmatrix} 1 & 0 & 0 \\ 0 & 4 & 0 \\ 0 & 0 & 1 \end{bmatrix}$$

This elementary matrix has been generated by multiplying row 2 of the 3×3 identity matrix by 4. Representing this elementary matrix by $E_{2(4)}$ we write:

$$E_{2(4)} \begin{bmatrix} a & b & c \\ d & e & f \\ g & h & i \end{bmatrix} = \begin{bmatrix} 1 & 0 & 0 \\ 0 & 4 & 0 \\ 0 & 0 & 1 \end{bmatrix} \begin{bmatrix} a & b & c \\ d & e & f \\ g & h & i \end{bmatrix}$$

which gives

$$\begin{bmatrix} a & b & c \\ 4d & 4e & 4f \\ g & h & i \end{bmatrix}$$

In general the elementary matrix which can be used to multiply row r of a given matrix by the scalar n is represented by $E_{r(n)}$. The following are some examples.

Worked example

7.9 Find the elementary matrices that may be used to achieve the following transformations.

(a) Transform the matrix

$$\begin{bmatrix} u & v \\ w & x \end{bmatrix} \quad \text{to} \quad \begin{bmatrix} 2u & 2v \\ w & x \end{bmatrix}$$

(b) Transform the matrix

$$\begin{bmatrix} 2 & 4 & 6 \\ 8 & 10 & 12 \\ 14 & 16 & 18 \end{bmatrix} \quad \text{to} \quad \begin{bmatrix} 2 & 4 & 6 \\ 8 & 10 & 12 \\ 7 & 8 & 9 \end{bmatrix}$$

Solution (a) In this case row 1 of the matrix has been multiplied by 2. This transformation can be achieved by premultiplying the matrix by $E_{1(2)}$. We have

$$E_{1(2)} \begin{bmatrix} u & v \\ w & x \end{bmatrix} = \begin{bmatrix} 2 & 0 \\ 0 & 1 \end{bmatrix} \begin{bmatrix} u & v \\ w & x \end{bmatrix}$$

which gives

$$\begin{bmatrix} 2u & 2v \\ w & x \end{bmatrix}$$

(b) In this case row 3 of the matrix has been multiplied by 1/2. This can be achieved by premultiplying the matrix by $E_{3(1/2)}$. In this case we have

$$E_{3(1/2)}\begin{bmatrix}2 & 4 & 6\\ 8 & 10 & 12\\ 14 & 16 & 18\end{bmatrix} = \begin{bmatrix}1 & 0 & 0\\ 0 & 1 & 0\\ 0 & 0 & \frac{1}{2}\end{bmatrix}\begin{bmatrix}2 & 4 & 6\\ 8 & 10 & 12\\ 14 & 16 & 18\end{bmatrix}$$

which gives

$$\begin{bmatrix}2 & 4 & 6\\ 8 & 10 & 12\\ 7 & 8 & 9\end{bmatrix}$$

Adding one row of a matrix to another

If we have the matrix

$$\begin{bmatrix}2 & 3 & 4\\ 5 & 6 & 7\\ 8 & 9 & 10\end{bmatrix}$$

the result of adding row 2 of this matrix to row 1 and using the result as the new row 1 is

$$\begin{bmatrix}7 & 9 & 11\\ 5 & 6 & 7\\ 8 & 9 & 10\end{bmatrix}$$

This transformation can be achieved by premultiplying the matrix by the elementary matrix

$$\begin{bmatrix}1 & 1 & 0\\ 0 & 1 & 0\\ 0 & 0 & 1\end{bmatrix}$$

This matrix is obtained by increasing the values of row 1 of the 3×3 identity matrix by the corresponding values in row 2. We represent this matrix by E_{1+2} and we write

$$E_{1+2}\begin{bmatrix}2 & 3 & 4\\ 5 & 6 & 7\\ 8 & 9 & 10\end{bmatrix} = \begin{bmatrix}1 & 1 & 0\\ 0 & 1 & 0\\ 0 & 0 & 1\end{bmatrix}\begin{bmatrix}2 & 3 & 4\\ 5 & 6 & 7\\ 8 & 9 & 10\end{bmatrix}$$

which gives

$$\begin{bmatrix}7 & 9 & 11\\ 5 & 6 & 7\\ 8 & 9 & 10\end{bmatrix}$$

In general the elementary matrix which can be used to increase the values of the elements of row $r1$ of a given matrix by adding to them the values of the corresponding elements in row $r2$ is represented by E_{r1+r2}. Here are some more examples of this type of transformation.

Worked example

7.10 Find the elementary matrices which can be used to achieve the following transformations:

(a) From $\begin{bmatrix} 1 & 2 & 3 \\ 4 & 5 & 6 \\ 7 & 8 & 9 \end{bmatrix}$ to $\begin{bmatrix} 8 & 10 & 12 \\ 4 & 5 & 6 \\ 7 & 8 & 9 \end{bmatrix}$

(b) From $\begin{bmatrix} x & y & z \\ 2x & y & 2z \\ x^2 & y^2 & z^2 \end{bmatrix}$ to $\begin{bmatrix} x & y & z \\ 2x & y & 2z \\ x^2+x & y^2+y & z^2+z \end{bmatrix}$

Solution (a) In this case we wish to add row 3 of the matrix to row 1. This can be achieved by using the elementary matrix E_{3+1}. We get

$$E_{1+3}\begin{bmatrix} 1 & 2 & 3 \\ 4 & 5 & 6 \\ 7 & 8 & 9 \end{bmatrix} = \begin{bmatrix} 1 & 0 & 1 \\ 0 & 1 & 0 \\ 0 & 0 & 1 \end{bmatrix}\begin{bmatrix} 1 & 2 & 3 \\ 4 & 5 & 6 \\ 7 & 8 & 9 \end{bmatrix}$$

which gives

$$\begin{bmatrix} 8 & 10 & 12 \\ 4 & 5 & 6 \\ 7 & 8 & 9 \end{bmatrix}$$

(b) In this case we wish to add row 1 of the matrix to row 3. This can be achieved by using the elementary matrix E_{3+1}. We get

$$E_{3+1}\begin{bmatrix} x & y & z \\ 2x & y & 2z \\ x^2 & y^2 & z^2 \end{bmatrix} = \begin{bmatrix} 1 & 0 & 0 \\ 0 & 1 & 0 \\ 1 & 0 & 1 \end{bmatrix}\begin{bmatrix} x & y & z \\ 2x & y & 2z \\ x^2 & y^2 & z^2 \end{bmatrix}$$

which gives

$$\begin{bmatrix} x & y & z \\ 2x & y & 2z \\ x^2+x & y^2+y & z^2+z \end{bmatrix}$$

Exchanging the rows of a matrix

If we exchange row 3 with row 1 in the matrix

$$\begin{bmatrix} a & b & c \\ d & e & f \\ g & h & i \end{bmatrix}$$

we will obtain the matrix

$$\begin{bmatrix} g & h & i \\ d & e & f \\ a & b & c \end{bmatrix}$$

This transformation may be achieved by premultiplying the matrix by the elementary matrix

$$\begin{bmatrix} 0 & 0 & 1 \\ 0 & 1 & 0 \\ 1 & 0 & 0 \end{bmatrix}$$

which is obtained by interchanging rows 3 and 1 of the identity matrix. We represent this matrix by $E_{1,3}$ and we write

$$E_{1,3}\begin{bmatrix} a & b & c \\ d & e & f \\ g & h & i \end{bmatrix} = \begin{bmatrix} 0 & 0 & 1 \\ 0 & 1 & 0 \\ 1 & 0 & 0 \end{bmatrix}\begin{bmatrix} a & b & c \\ d & e & f \\ g & h & i \end{bmatrix}$$

which gives

$$\begin{bmatrix} g & h & i \\ d & e & f \\ a & b & c \end{bmatrix}$$

Here are some more examples.

Worked example

7.11 Find the elementary matrices that may be used to achieve the following transformations.

(a) From $\begin{bmatrix} a & b & c \\ d & e & f \\ g & h & i \end{bmatrix}$ to $\begin{bmatrix} a & b & c \\ g & h & i \\ d & e & f \end{bmatrix}$

(b) From $\begin{bmatrix} 1 & 2 \\ 3 & 4 \end{bmatrix}$ to $\begin{bmatrix} 3 & 4 \\ 1 & 2 \end{bmatrix}$

Solution (a) In this case we wish to exchange rows 3 and 2 of the matrix. This can be achieved using the elementary matrix $E_{3,2}$. We get

$$E_{3,2}\begin{bmatrix} a & b & c \\ d & e & f \\ g & h & i \end{bmatrix} = \begin{bmatrix} 1 & 0 & 0 \\ 0 & 0 & 1 \\ 0 & 1 & 0 \end{bmatrix}\begin{bmatrix} a & b & c \\ d & e & f \\ g & h & i \end{bmatrix}$$

which gives

$$\begin{bmatrix} a & b & c \\ g & h & i \\ d & e & f \end{bmatrix}$$

(b) In this case we wish to exchange rows 1 and 2 of the matrix. Here we use $E_{1,2}$ to get

$$E_{1,2}\begin{bmatrix} 1 & 2 \\ 3 & 4 \end{bmatrix} = \begin{bmatrix} 0 & 1 \\ 1 & 0 \end{bmatrix}\begin{bmatrix} 1 & 2 \\ 3 & 4 \end{bmatrix}$$

which gives

$$\begin{bmatrix} 3 & 4 \\ 1 & 2 \end{bmatrix}$$

Composite transformations

A transformation of a matrix which involves more than one elementary row operation we will call a **composite transformation**.

It is possible to achieve a composite transformation of a given matrix by performing a sequence of premultiplications by elementary matrices. For example, the transformation of the matrix

$$\begin{bmatrix} a & b & c \\ d & e & f \\ g & h & i \end{bmatrix} \quad \text{into} \quad \begin{bmatrix} a+2g & b+2h & c+2i \\ d & e & f \\ g & h & i \end{bmatrix}$$

would involve adding twice the value of row 3 to row 1. We know that the process of adding row 3 of the matrix to row 1 can be achieved by premultiplying the matrix by the elementary matrix E_{1+3}. To achieve the desired transformation we could perform this process twice, that is to say premultiply the matrix by E_{1+3} and by E_{1+3} again. We may represent this transformation as

$$E_{1+3}E_{1+3}\begin{bmatrix} a & b & c \\ d & e & f \\ g & h & i \end{bmatrix}$$

Performing the multiplication $E_{1+3}\ E_{1+3}$ we get

$$E_{1+3}E_{1+3} = \begin{bmatrix} 1 & 0 & 1 \\ 0 & 1 & 0 \\ 0 & 0 & 1 \end{bmatrix}\begin{bmatrix} 1 & 0 & 1 \\ 0 & 1 & 0 \\ 0 & 0 & 1 \end{bmatrix}$$

giving

$$E_{1+3}E_{1+3} = \begin{bmatrix} 1 & 0 & 2 \\ 0 & 1 & 0 \\ 0 & 0 & 1 \end{bmatrix}$$

We will represent this new elementary matrix by $E_{1+3(2)}$. The desired transformation can now be performed as:

$$E_{1+3(2)}\begin{bmatrix} a & b & c \\ d & e & f \\ g & h & i \end{bmatrix} = \begin{bmatrix} 1 & 0 & 2 \\ 0 & 1 & 0 \\ 0 & 0 & 1 \end{bmatrix}\begin{bmatrix} a & b & c \\ d & e & f \\ g & h & i \end{bmatrix}$$

which gives the desired result of

$$\begin{bmatrix} a+2g & b+2h & c+2i \\ d & e & f \\ g & h & i \end{bmatrix}$$

More generally if we wished to add n times row 3 of the matrix to row 1 we could achieve this transformation by premultiplying the matrix by $E_{1+3(n)}$ which is

$$\begin{bmatrix} 1 & 0 & n \\ 0 & 1 & 0 \\ 0 & 0 & 1 \end{bmatrix}$$

Worked examples

7.12 Find the elementary matrix which can be used to perform the transformation of the matrix

$$\begin{bmatrix} 1 & 2 & 3 \\ 4 & 5 & 6 \\ 7 & 8 & 9 \end{bmatrix} \quad \text{to the matrix} \quad \begin{bmatrix} 1 & 2 & 3 \\ 4 & 5 & 6 \\ 6 & 6 & 6 \end{bmatrix}$$

Solution This transformation may be achieved by subtracting row 1 of the matrix from row 3. An equivalent process is to add -1 times row 1 to row 3. From above we know that we may perform this transformation by premultiplying the matrix by $E_{3+1(-1)}$. This gives

$$E_{3+1(-1)} \begin{bmatrix} 1 & 2 & 3 \\ 4 & 5 & 6 \\ 7 & 8 & 9 \end{bmatrix} = \begin{bmatrix} 1 & 0 & 0 \\ 0 & 1 & 0 \\ -1 & 0 & 1 \end{bmatrix} \begin{bmatrix} 1 & 2 & 3 \\ 4 & 5 & 6 \\ 7 & 8 & 9 \end{bmatrix}$$

which gives

$$\begin{bmatrix} 1 & 2 & 3 \\ 4 & 5 & 6 \\ 6 & 6 & 6 \end{bmatrix}$$

7.13 Identify the sequence of row operations that must be performed to transform the matrix

$$\begin{bmatrix} 1 & 2 & 3 \\ 4 & 5 & 6 \\ 7 & 8 & 9 \end{bmatrix} \quad \text{into the matrix} \quad \begin{bmatrix} 6 & 9 & 12 \\ 1 & 2 & 3 \\ 7 & 8 & 9 \end{bmatrix}$$

Calculate the matrix which may be used to achieve the transformation.

Solution The transformation of the matrix

$$\begin{bmatrix} 1 & 2 & 3 \\ 4 & 5 & 6 \\ 7 & 8 & 9 \end{bmatrix} \quad \text{into the matrix} \quad \begin{bmatrix} 6 & 9 & 12 \\ 1 & 2 & 3 \\ 7 & 8 & 9 \end{bmatrix}$$

may be achieved by adding twice row 1 of the matrix to row 2 and then by exchanging rows 1 and 2. The elementary matrices which can be used to achieve each of these transformations are $E_{2+1(2)}$ and $E_{1,2}$. If we apply each of these matrices in turn then we will achieve the desired composite transformation. That is the transformation may be achieved by premultiplying the given matrix by $E_{1,2}E_{2+1(2)}$. Now we can calculate the value of $E_{1,2}E_{2+1(2)}$ as follows:

$$E_{1,2}\,E_{2+1(2)} = \begin{bmatrix} 0 & 1 & 0 \\ 1 & 0 & 0 \\ 0 & 0 & 1 \end{bmatrix} \begin{bmatrix} 1 & 0 & 0 \\ 2 & 1 & 0 \\ 0 & 0 & 1 \end{bmatrix}$$

which gives

$$\begin{bmatrix} 2 & 1 & 0 \\ 1 & 0 & 0 \\ 0 & 0 & 1 \end{bmatrix}$$

Applying this matrix we get

$$\begin{bmatrix} 2 & 1 & 0 \\ 1 & 0 & 0 \\ 0 & 0 & 1 \end{bmatrix}\begin{bmatrix} 1 & 2 & 3 \\ 4 & 5 & 6 \\ 7 & 8 & 9 \end{bmatrix}$$

which gives the desired result of

$$\begin{bmatrix} 6 & 9 & 12 \\ 1 & 2 & 3 \\ 7 & 8 & 9 \end{bmatrix}$$

Self-Assessment Questions 7.5

What would be the effect on some matrix $\boldsymbol{A}$ if it were postmultiplied by some elementary matrix $\boldsymbol{E}$? What can you say about the dimensions of $\boldsymbol{A}$ if it is possible to perform this multiplication?

Exercise 7.5

1. Find the elementary matrices which may be used to achieve the following transformations.

(a) From $\begin{bmatrix} 1 & 2 & 3 \\ 4 & 5 & 6 \\ 7 & 8 & 9 \end{bmatrix}$ to $\begin{bmatrix} 5 & 10 & 15 \\ 4 & 5 & 6 \\ 7 & 8 & 9 \end{bmatrix}$

(b) From $\begin{bmatrix} a & b & c \\ d & e & f \\ g & h & i \end{bmatrix}$ to $\begin{bmatrix} a & b & c \\ d^2 & de & df \\ g & h & i \end{bmatrix}$

(c) From $\begin{bmatrix} 2 & 3 & 4 \\ 5 & 6 & 7 \\ 8 & 9 & 10 \end{bmatrix}$ to $\begin{bmatrix} 2 & 3 & 4 \\ 7 & 9 & 11 \\ 8 & 9 & 10 \end{bmatrix}$

(d) From $\begin{bmatrix} x \\ y \\ z \end{bmatrix}$ to $\begin{bmatrix} x \\ y \\ y+z \end{bmatrix}$

(e) From $\begin{bmatrix} 2 & 3 & 4 & 5 \\ 5 & 6 & 7 & 9 \\ 1 & 5 & 4 & 3 \\ 8 & 9 & 0 & 4 \end{bmatrix}$ to $\begin{bmatrix} 2 & 3 & 4 & 5 \\ 5 & 6 & 7 & 9 \\ 1 & 5 & 4 & 3 \\ 10 & 12 & 4 & 9 \end{bmatrix}$

(f) From $\begin{bmatrix} a & b \\ c & d \\ e & f \end{bmatrix}$ to $\begin{bmatrix} e & f \\ c & d \\ a & b \end{bmatrix}$

(g) From $\begin{bmatrix} 1 & 2 & 3 \\ 4 & 5 & 6 \\ 7 & 8 & 9 \end{bmatrix}$ to $\begin{bmatrix} 1 & 2 & 3 \\ 7 & 8 & 9 \\ 4 & 5 & 6 \end{bmatrix}$

2. For each of the transformations shown below identify the sequence of elementary row operations that may be performed to achieve the transformation and calculate the matrix which can be used to accomplish the transformation.

(a) From $\begin{bmatrix} a & b & c \\ d & e & f \\ g & h & i \end{bmatrix}$

to $\begin{bmatrix} a & b & c \\ g-2a & h-2b & i-2c \\ d & e & f \end{bmatrix}$

(b) From $\begin{bmatrix} 2 & 3 & 4 \\ 5 & 6 & 7 \\ 8 & 9 & 10 \end{bmatrix}$ to $\begin{bmatrix} 2 & 3 & 4 \\ 3 & 3 & 3 \\ 4 & 3 & 2 \end{bmatrix}$

(c) From $\begin{bmatrix} 1 & 2 & 3 \\ 6 & 9 & 12 \\ 4 & 6 & 8 \end{bmatrix}$ to $\begin{bmatrix} 1 & 2 & 3 \\ 2 & 3 & 4 \\ 1 & 0 & -1 \end{bmatrix}$

7.6 The inverse of a matrix

If it is possible to find a matrix $\boldsymbol{B}$ for some given matrix $\boldsymbol{A}$ so that the equations $\boldsymbol{BA} = \boldsymbol{I}$ and $\boldsymbol{AB} = \boldsymbol{I}$ ($\boldsymbol{I}$ is the identity matrix) hold, then the matrix $\boldsymbol{A}$ is said to be **non-singular** and the matrix $\boldsymbol{B}$ is said to be the **inverse** of $\boldsymbol{A}$. If the inverse of the matrix $\boldsymbol{A}$ does exist then it is usually represented by $\boldsymbol{A}^{-1}$, and so the equations above would be written as

$$\boldsymbol{A}^{-1}\boldsymbol{A} = \boldsymbol{I} \quad \text{and} \quad \boldsymbol{AA}^{-1} = \boldsymbol{I}$$

If it is not possible to find the inverse of a particular matrix then that matrix is said to be **singular**.

KEY POINT

For a matrix to be non-singular the following three conditions must be met.

1 The matrix must be square.
2 There should be no column of the matrix which consists only of zeros.
3 It should not be possible to generate any row of the matrix from the other rows using only elementary row operations.

Calculating the inverse of a non-singular matrix

Consider the non-singular matrix $\boldsymbol{A}$ which is

$$\begin{bmatrix} 3 & 6 \\ 6 & 14 \end{bmatrix}$$

Since this matrix is non-singular it is possible to transform it into the identity matrix by premultiplying it by a sequence of elementary matrices. The steps in this calculation are shown below. The right-hand column contains the elementary matrices used along with a description of their effects.

$$\begin{bmatrix} 3 & 6 \\ 6 & 14 \end{bmatrix}$$

$\Downarrow$ $\qquad E_{1(1/3)}$

Multiply row 1 by 1/3

$$\begin{bmatrix} 1 & 2 \\ 6 & 14 \end{bmatrix}$$

$\Downarrow$ $\qquad E_{2+1(-6)}$

Add -6 times row 1 to row 2

$$\begin{bmatrix} 1 & 2 \\ 0 & 2 \end{bmatrix}$$

$$\Downarrow \quad E_{1+2(-1)}$$

Add – 1 times row 2 to row 1

$$\begin{bmatrix} 1 & 0 \\ 0 & 2 \end{bmatrix}$$

$$\Downarrow \quad E_{2(1/2)}$$

Multiply row 2 by 1/2

$$\begin{bmatrix} 1 & 0 \\ 0 & 1 \end{bmatrix}$$

Since the sequence of premultiplications above transform $\boldsymbol{A}$ into $\boldsymbol{I}$ we can write

$$E_{2(1/2)}E_{1+2(-1)}E_{2+1(-6)}E_{1(1/3)}\boldsymbol{A} = \boldsymbol{I}$$

The equation defining the inverse of a matrix states that

$$\boldsymbol{A}^{-1}\boldsymbol{A} = \boldsymbol{I}$$

Hence we can write

$$\boldsymbol{A}^{-1} = E_{2(1/2)}E_{1+2(-1)}E_{2+1(-6)}E_{1(1/3)}$$

Evaluating the right-hand side of this equation we get

$$\begin{bmatrix} 1 & 0 \\ 0 & 1/2 \end{bmatrix} \begin{bmatrix} 1 & -1 \\ 0 & 1 \end{bmatrix} \begin{bmatrix} 1 & 0 \\ -6 & 1 \end{bmatrix} \begin{bmatrix} 1/3 & 0 \\ 0 & 1 \end{bmatrix}$$

$$= \begin{bmatrix} 1 & 0 \\ 0 & 1/2 \end{bmatrix} \begin{bmatrix} 1 & -1 \\ 0 & 1 \end{bmatrix} \begin{bmatrix} 1/3 & 0 \\ -2 & 1 \end{bmatrix}$$

$$= \begin{bmatrix} 1 & 0 \\ 0 & 1/2 \end{bmatrix} \begin{bmatrix} 7/3 & -1 \\ -2 & 1 \end{bmatrix} = \begin{bmatrix} 7/3 & -1 \\ -1 & 1/2 \end{bmatrix}$$

We have calculated the inverse of the matrix $\boldsymbol{A}$ and we can write

$$\boldsymbol{A}^{-1} = \begin{bmatrix} 7/3 & -1 \\ -1 & 1/2 \end{bmatrix}$$

To check that this is correct we calculate

$$\boldsymbol{A}^{-1}\boldsymbol{A}$$

which is

$$\begin{bmatrix} 7/3 & -1 \\ -1 & 1/2 \end{bmatrix} \begin{bmatrix} 3 & 6 \\ 6 & 14 \end{bmatrix}$$

which gives

$$\begin{bmatrix} (7/3 \times 3) - 6 & (7/3 \times 6) - 14 \\ -3 + (1/2 \times 6) & -6 + (1/2 \times 14) \end{bmatrix}$$

which simplifies to

$$\begin{bmatrix} 1 & 0 \\ 0 & 1 \end{bmatrix}$$

So we see that the calculation of the inverse is correct.

The following are two somewhat longer examples.

Worked examples

7.14 Calculate the inverse of the matrix

$$\begin{bmatrix} 1 & 0 & 2 \\ 2 & 1 & 0 \\ 0 & 2 & 1 \end{bmatrix}$$

Solution

$$\begin{bmatrix} 1 & 0 & 2 \\ 2 & 1 & 0 \\ 0 & 2 & 1 \end{bmatrix}$$

$$\Downarrow \quad E_{1+3(-2)}$$

Add -2 times row 3 to row 1

$$\begin{bmatrix} 1 & -4 & 0 \\ 2 & 1 & 0 \\ 0 & 2 & 1 \end{bmatrix}$$

$$\Downarrow \quad E_{1+2(4)}$$

Add 4 times row 2 to row 1

$$\begin{bmatrix} 9 & 0 & 0 \\ 2 & 1 & 0 \\ 0 & 2 & 1 \end{bmatrix}$$

$$\Downarrow \quad E_{1(1/9)}$$

Multiply row 1 by 1/9

$$\begin{bmatrix} 1 & 0 & 0 \\ 2 & 1 & 0 \\ 0 & 2 & 1 \end{bmatrix}$$

$$\Downarrow \quad E_{2+1(-2)}$$

Add -2 times row 1 to row 2

$$\begin{bmatrix} 1 & 0 & 0 \\ 0 & 1 & 0 \\ 0 & 2 & 1 \end{bmatrix}$$

$$\Downarrow \quad E_{3+2(-2)}$$

Add -2 times row 2 to row 3

$$\begin{bmatrix} 1 & 0 & 0 \\ 0 & 1 & 0 \\ 0 & 0 & 1 \end{bmatrix}$$

Multiplying the elementary matrices together we get

$$E_{3+2(-2)}E_{2+1(-2)}E_{1(1/9)}E_{1+2(4)}E_{1+3(-2)}$$

This expression evaluates to

$$\begin{bmatrix}1 & 0 & 0\\ 0 & 1 & 0\\ 0 & -2 & 1\end{bmatrix}\begin{bmatrix}1 & 0 & 0\\ -2 & 1 & 0\\ 0 & 0 & 1\end{bmatrix}\begin{bmatrix}1/9 & 0 & 0\\ 0 & 1 & 0\\ 0 & 0 & 1\end{bmatrix}\begin{bmatrix}1 & 4 & 0\\ 0 & 1 & 0\\ 0 & 0 & 1\end{bmatrix}\begin{bmatrix}1 & 0 & -2\\ 0 & 1 & 0\\ 0 & 0 & 1\end{bmatrix}$$

$$= \begin{bmatrix}1 & 0 & 0\\ 0 & 1 & 0\\ 0 & -2 & 1\end{bmatrix}\begin{bmatrix}1 & 0 & 0\\ -2 & 1 & 0\\ 0 & 0 & 1\end{bmatrix}\begin{bmatrix}1/9 & 0 & 0\\ 0 & 1 & 0\\ 0 & 0 & 1\end{bmatrix}\begin{bmatrix}1 & 4 & -2\\ 0 & 1 & 0\\ 0 & 0 & 1\end{bmatrix}$$

$$= \begin{bmatrix}1 & 0 & 0\\ 0 & 1 & 0\\ 0 & -2 & 1\end{bmatrix}\begin{bmatrix}1 & 0 & 0\\ -2 & 1 & 0\\ 0 & 0 & 1\end{bmatrix}\begin{bmatrix}1/9 & 4/9 & -2/9\\ 0 & 1 & 0\\ 0 & 0 & 1\end{bmatrix}$$

$$= \begin{bmatrix}1 & 0 & 0\\ 0 & 1 & 0\\ 0 & -2 & 1\end{bmatrix}\begin{bmatrix}1/9 & 4/9 & -2/9\\ -2/9 & 1/9 & 4/9\\ 0 & 0 & 1\end{bmatrix}$$

$$= \begin{bmatrix}1/9 & 4/9 & -2/9\\ -2/9 & 1/9 & 4/9\\ 4/9 & -2/9 & 1/9\end{bmatrix}$$

The above solution is rather lengthy; we may make it more compact by considering the following facts. If a matrix $\boldsymbol{A}$ can be transformed into the identity matrix $\boldsymbol{I}$ by premultiplying by a sequence of elementary matrices $E_1 \ldots E_n$ then we can write

$$E_n \ldots E_1 \boldsymbol{A} = \boldsymbol{I}$$

From this equation it follows that

$$E_n \ldots E_1 = \boldsymbol{A}^{-1}$$

which can be written as

$$E_n \ldots E_1 \boldsymbol{I} = \boldsymbol{A}^{-1}$$

The two equations

$$E_n \ldots E_1 \boldsymbol{A} = \boldsymbol{I}$$

and

$$E_n \ldots E_1 \boldsymbol{I} = \boldsymbol{A}^{-1}$$

tell us that the sequence of elementary row operations that transform the matrix $\boldsymbol{A}$ into the identity matrix $\boldsymbol{I}$ is also the sequence that will transform the identity matrix $\boldsymbol{I}$ into the inverse of $\boldsymbol{A}$, namely $\boldsymbol{A}^{-1}$. We can use this fact to reduce the length of the above solution. We calculate the inverse of the matrix $\boldsymbol{A}$ progressively, by performing the same sequence of elementary row operations on the identity matrix $\boldsymbol{I}$ as we perform on the matrix $\boldsymbol{A}$ in order to turn $\boldsymbol{A}$ into $\boldsymbol{I}$.

Here is the more compact solution:

$$\begin{bmatrix} 1 & 0 & 0 \\ 0 & 1 & 0 \\ 0 & 0 & 1 \end{bmatrix} \begin{bmatrix} 1 & 0 & 2 \\ 2 & 1 & 0 \\ 0 & 2 & 1 \end{bmatrix}$$

Add -2 times row 3 to row 1 ($E_{1+3(-2)}$)

$$\begin{bmatrix} 1 & 0 & -2 \\ 0 & 1 & 0 \\ 0 & 0 & 1 \end{bmatrix} \begin{bmatrix} 1 & -4 & 0 \\ 2 & 1 & 0 \\ 0 & 2 & 1 \end{bmatrix}$$

Add 4 times row 2 to row 1 ($E_{1+2(4)}$)

$$\begin{bmatrix} 1 & 4 & -2 \\ 0 & 1 & 0 \\ 0 & 0 & 1 \end{bmatrix} \begin{bmatrix} 9 & 0 & 0 \\ 2 & 1 & 0 \\ 0 & 2 & 1 \end{bmatrix}$$

Multiply row 1 by 1/9 ($E_{1(1/9)}$)

$$\begin{bmatrix} 1/9 & 4/9 & -2/9 \\ 0 & 1 & 0 \\ 0 & 0 & 1 \end{bmatrix} \begin{bmatrix} 1 & 0 & 0 \\ 2 & 1 & 0 \\ 0 & 2 & 1 \end{bmatrix}$$

Add -2 times row 1 to row 2 ($E_{2+1(-2)}$)

$$\begin{bmatrix} 1/9 & 4/9 & -2/9 \\ -2/9 & 1/9 & 4/9 \\ 0 & 0 & 1 \end{bmatrix} \begin{bmatrix} 1 & 0 & 0 \\ 0 & 1 & 0 \\ 0 & 2 & 1 \end{bmatrix}$$

Add -2 times row 2 to row 3 ($E_{3+2(-2)}$)

$$\begin{bmatrix} 1/9 & 4/9 & -2/9 \\ -2/9 & 1/9 & 4/9 \\ 4/9 & -2/9 & 1/9 \end{bmatrix} \begin{bmatrix} 1 & 0 & 0 \\ 0 & 1 & 0 \\ 0 & 0 & 1 \end{bmatrix}$$

Again the inverse of the given matrix has been calculated to be

$$\begin{bmatrix} 1/9 & 4/9 & -2/9 \\ -2/9 & 1/9 & 4/9 \\ 4/9 & -2/9 & 1/9 \end{bmatrix}$$

7.15 Calculate the inverse of the following matrix and show that the result is correct.

$$\begin{bmatrix} 1 & 0 & 1 \\ 2 & 2 & 0 \\ 0 & 1 & 3 \end{bmatrix}$$

Solution The operation involved in each step is indicated by the appropriate elementary matrix

$$\begin{bmatrix} 1 & 0 & 0 \\ 0 & 1 & 0 \\ 0 & 0 & 1 \end{bmatrix} \begin{bmatrix} 1 & 0 & 1 \\ 2 & 2 & 0 \\ 0 & 1 & 3 \end{bmatrix}$$

$E_{2(1/2)}$

$$\begin{bmatrix} 1 & 0 & 0 \\ 0 & 1/2 & 0 \\ 0 & 0 & 1 \end{bmatrix} \begin{bmatrix} 1 & 0 & 1 \\ 1 & 1 & 0 \\ 0 & 1 & 3 \end{bmatrix}$$

$E_{2+1(-1)}$

$$\begin{bmatrix} 1 & 0 & 0 \\ -1 & 1/2 & 0 \\ 0 & 0 & 1 \end{bmatrix} \begin{bmatrix} 1 & 0 & 1 \\ 0 & 1 & -1 \\ 0 & 1 & 3 \end{bmatrix}$$

$E_{3+2(-1)}$

$$\begin{bmatrix} 1 & 0 & 0 \\ -1 & 1/2 & 0 \\ 1 & -1/2 & 1 \end{bmatrix} \begin{bmatrix} 1 & 0 & 1 \\ 0 & 1 & -1 \\ 0 & 0 & 4 \end{bmatrix}$$

$E_{3(1/4)}$

$$\begin{bmatrix} 1 & 0 & 0 \\ -1 & 1/2 & 0 \\ 1/4 & -1/8 & 1/4 \end{bmatrix} \begin{bmatrix} 1 & 0 & 1 \\ 0 & 1 & -1 \\ 0 & 0 & 1 \end{bmatrix}$$

$E_{1+3(-1)}$

$$\begin{bmatrix} 3/4 & 1/8 & -1/4 \\ -1 & 1/2 & 0 \\ 1/4 & -1/8 & 1/4 \end{bmatrix} \begin{bmatrix} 1 & 0 & 0 \\ 0 & 1 & -1 \\ 0 & 0 & 1 \end{bmatrix}$$

E_{2+3}

$$\begin{bmatrix} 3/4 & 1/8 & -1/4 \\ -3/4 & 3/8 & 1/4 \\ 1/4 & -1/8 & 1/4 \end{bmatrix} \begin{bmatrix} 1 & 0 & 0 \\ 0 & 1 & 0 \\ 0 & 0 & 1 \end{bmatrix}$$

Checking the calculation:

$$\begin{bmatrix} 3/4 & 1/8 & -1/4 \\ -3/4 & 3/8 & 1/4 \\ 1/4 & -1/8 & 1/4 \end{bmatrix} \begin{bmatrix} 1 & 0 & 1 \\ 2 & 2 & 0 \\ 0 & 1 & 3 \end{bmatrix}$$

gives

$$\begin{bmatrix} (3/4 + 2/8) & (2/8 - 1/4) & (3/4 - 3/4) \\ (-3/4 + 6/8) & (6/8 + 1/4) & (-3/4 + 3/4) \\ (1/4 - 2/8) & (-2/8 + 1/4) & (1/4 + 3/4) \end{bmatrix}$$

which is the desired result of

$$\begin{bmatrix} 1 & 0 & 0 \\ 0 & 1 & 0 \\ 0 & 0 & 1 \end{bmatrix}$$

Self-Assessment Question 7.6

What can you say about the matrix A if $A^{-1} = I$?

Exercise 7.6

1. Determine why each of the following matrices is singular.

(a) $\begin{bmatrix} 1 & 6 & 2 \\ 3 & 4 & 5 \end{bmatrix}$ (b) $\begin{bmatrix} 1 & 2 & 0 & 4 \\ 7 & 8 & 0 & 3 \\ 5 & 3 & 0 & 9 \\ 3 & 4 & 0 & 6 \end{bmatrix}$

(c) $\begin{bmatrix} 2 & 5 & 6 & 3 \\ 4 & 7 & 2 & 9 \\ 2 & 2 & -4 & 6 \\ 1 & 3 & 5 & 9 \end{bmatrix}$ (d) $\begin{bmatrix} 3 & 2 & 4 \\ 4 & 5 & 9 \\ 9 & 6 & 12 \end{bmatrix}$

2. Calculate the inverse of the following matrices.

(a) $\begin{bmatrix} 2 & 3 \\ 1 & -1 \end{bmatrix}$ (b) $\begin{bmatrix} 3 & 7 \\ 6 & 12 \end{bmatrix}$ (c) $\begin{bmatrix} 4 & 1 \\ 1 & 3 \end{bmatrix}$

3. Calculate the inverse of the following matrices and check that your calculation is correct.

(a) $\begin{bmatrix} 2 & 3 \\ -1 & 2 \end{bmatrix}$ (b) $\begin{bmatrix} 2 & 1 & 4 \\ 3 & 5 & 1 \\ 1 & 2 & 0 \end{bmatrix}$

(c) $\begin{bmatrix} 2 & -1 & 0 \\ 1 & 2 & -1 \\ -1 & -1 & 2 \end{bmatrix}$

7.7 Solving systems of equations

Consider the following system of two equations:

$$2x + y = 10$$
$$3x + 2y = 17$$

We may represent this pair of equations as a single equation containing two matrices, namely,

$$\begin{bmatrix} 2x + y \\ 3x + 2y \end{bmatrix} = \begin{bmatrix} 10 \\ 17 \end{bmatrix}$$

The fact that the system of two equations and the single equation involving matrices are equivalent relies upon the fact that for two matrices to be equal their corresponding elements must be equal.

The left-hand matrix in the above equation may be written as the product of two matrices, one containing only numeric values and the other containing only the variables x and y. This product is

$$\begin{bmatrix} 2 & 1 \\ 3 & 2 \end{bmatrix} \begin{bmatrix} x \\ y \end{bmatrix}$$

So the original system of two equations may be represented as

$$\begin{bmatrix} 2 & 1 \\ 3 & 2 \end{bmatrix}\begin{bmatrix} x \\ y \end{bmatrix} = \begin{bmatrix} 10 \\ 17 \end{bmatrix}$$

If we represent the matrices that contain the numeric values as $\boldsymbol{A}$ and $\boldsymbol{R}$ then we may write

$$\boldsymbol{A}\begin{bmatrix} x \\ y \end{bmatrix} = \boldsymbol{R}$$

If $\boldsymbol{A}$ is non-singular then it will have an inverse $\boldsymbol{A}^{-1}$. If we premultiply both sides of the above equation by this inverse we get

$$\boldsymbol{A}^{-1}\boldsymbol{A}\begin{bmatrix} x \\ y \end{bmatrix} = \boldsymbol{A}^{-1}\boldsymbol{R}$$

This gives $\boldsymbol{I}\begin{bmatrix} x \\ y \end{bmatrix} = \boldsymbol{A}^{-1}\boldsymbol{R}$ which is just $\begin{bmatrix} x \\ y \end{bmatrix} = \boldsymbol{A}^{-1}\boldsymbol{R}$

Since $\boldsymbol{A}^{-1}$ and $\boldsymbol{R}$ contain only numeric values we may evaluate $\boldsymbol{A}^{-1}\boldsymbol{R}$ and hence determine the values of the variables x and y that satisfy the given pair of equations. Firstly we need to calculate $\boldsymbol{A}^{-1}$; this we can do as follows:

$$\begin{bmatrix} 1 & 0 \\ 0 & 1 \end{bmatrix}\begin{bmatrix} 2 & 1 \\ 3 & 2 \end{bmatrix}$$

$E_{1,2}$

$$\begin{bmatrix} 0 & 1 \\ 1 & 0 \end{bmatrix}\begin{bmatrix} 3 & 2 \\ 2 & 1 \end{bmatrix}$$

$E_{1+2(-2)}$

$$\begin{bmatrix} -2 & 1 \\ 1 & 0 \end{bmatrix}\begin{bmatrix} -1 & 0 \\ 2 & 1 \end{bmatrix}$$

$E_{1(-1)}$

$$\begin{bmatrix} 2 & -1 \\ 1 & 0 \end{bmatrix}\begin{bmatrix} 1 & 0 \\ 2 & 1 \end{bmatrix}$$

$E_{2+1(-2)}$

$$\begin{bmatrix} 2 & -1 \\ -3 & 2 \end{bmatrix}\begin{bmatrix} 1 & 0 \\ 0 & 1 \end{bmatrix}$$

So the value of $\boldsymbol{A}^{-1}$ is

$$\begin{bmatrix} 2 & -1 \\ -3 & 2 \end{bmatrix}$$

Hence $\begin{bmatrix} x \\ y \end{bmatrix} = \boldsymbol{A}^{-1}\boldsymbol{R}$ becomes $\begin{bmatrix} x \\ y \end{bmatrix} = \begin{bmatrix} 2 & -1 \\ -3 & 2 \end{bmatrix}\begin{bmatrix} 10 \\ 17 \end{bmatrix}$

which gives

$$\begin{bmatrix} x \\ y \end{bmatrix} = \begin{bmatrix} 3 \\ 4 \end{bmatrix}$$

Hence the given system of equations holds when x is 3 and y is 4.

Here is another example.

Worked example

7.16 Solve the following system of equations:

$$\begin{aligned} x - y + z &= 4 \\ 2x + y - 3z &= 5 \\ x + y + z &= 2 \end{aligned}$$

Solution The system of equations in matrix form is

$$\begin{bmatrix} 1 & -1 & 1 \\ 2 & 1 & -3 \\ 1 & 1 & 1 \end{bmatrix} \begin{bmatrix} x \\ y \\ z \end{bmatrix} = \begin{bmatrix} 4 \\ 5 \\ 2 \end{bmatrix}$$

Finding the inverse of the left-hand matrix we get

$$\begin{bmatrix} 1 & -1 & 1 \\ 2 & 1 & -3 \\ 1 & 1 & 1 \end{bmatrix} \begin{bmatrix} 1 & 0 & 0 \\ 0 & 1 & 0 \\ 0 & 0 & 1 \end{bmatrix}$$

E_{3+1}, $E_{3(1/2)}$

$$\begin{bmatrix} 1 & -1 & 1 \\ 2 & 1 & -3 \\ 0 & 1 & 0 \end{bmatrix} \begin{bmatrix} 1 & 0 & 0 \\ 0 & 1 & 0 \\ -1/2 & 0 & 1/2 \end{bmatrix}$$

$E_{2+1(-2)}$

$$\begin{bmatrix} 1 & -1 & 1 \\ 0 & 3 & -5 \\ 0 & 1 & 0 \end{bmatrix} \begin{bmatrix} 1 & 0 & 0 \\ -2 & 1 & 0 \\ -1/2 & 0 & 1/2 \end{bmatrix}$$

$E_{2+3(-3)}$, $E_{2(-1/5)}$

$$\begin{bmatrix} 1 & -1 & 1 \\ 0 & 0 & 1 \\ 0 & 1 & 0 \end{bmatrix} \begin{bmatrix} 1 & 0 & 0 \\ 1/10 & -1/5 & 3/10 \\ -1/2 & 0 & 1/2 \end{bmatrix}$$

$E_{1+2(-1)}$

$$\begin{bmatrix} 1 & -1 & 0 \\ 0 & 0 & 1 \\ 0 & 1 & 0 \end{bmatrix} \begin{bmatrix} 9/10 & 1/5 & -3/10 \\ 1/10 & -1/5 & 3/10 \\ -1/2 & 0 & 1/2 \end{bmatrix}$$

E_{1+3}, $E_{2,3}$

$$\begin{bmatrix} 1 & 0 & 0 \\ 0 & 1 & 0 \\ 0 & 0 & 1 \end{bmatrix} \begin{bmatrix} 4/10 & 1/5 & 1/5 \\ -1/2 & 0 & 1/2 \\ 1/10 & -1/5 & 3/10 \end{bmatrix}$$

The matrix equation may now be written as

$$\begin{bmatrix} x \\ y \\ z \end{bmatrix} = \begin{bmatrix} 4/10 & 1/5 & 1/5 \\ -1/2 & 0 & 1/2 \\ 1/10 & -1/5 & 3/10 \end{bmatrix} \begin{bmatrix} 4 \\ 5 \\ 2 \end{bmatrix}$$

which gives

$$\begin{bmatrix} x \\ y \\ z \end{bmatrix} = \begin{bmatrix} 3 \\ -1 \\ 0 \end{bmatrix}$$

Hence when the given system of equations hold x is 3, y is -1 and z is 0.

Exercise 7.7

1. Solve the following systems of equations:

(a) $2x + y = 10$
$3x + 2y = 17$

(b) $x + y - 2z = 4$
$2x - y + z = 3$
$3x + y - z = 2$

(c) $3a + 4b - 2c = 4$
$a + 2b + c = 1$
$-a + b + 2c = 2$

Test and Assignment Exercises 7

The first three questions in this exercise refer to the following matrices:

$$\boldsymbol{A} = \begin{bmatrix} 2 & 1 \\ 3 & 2 \end{bmatrix} \quad \boldsymbol{B} = \begin{bmatrix} 3 & 2 \\ 1 & 4 \end{bmatrix} \quad \boldsymbol{C} = \begin{bmatrix} a & 2b \\ 3c & d \end{bmatrix}$$

$$\boldsymbol{D} = \begin{bmatrix} a \\ b \end{bmatrix} \quad \boldsymbol{E} = [x \quad y] \quad \boldsymbol{F} = \begin{bmatrix} x \\ y \end{bmatrix}$$

$$\boldsymbol{G} = \begin{bmatrix} 2 & 4 & 1 \\ 3 & 2 & 4 \\ 1 & 3 & 2 \end{bmatrix} \quad \boldsymbol{H} = \begin{bmatrix} 2x & y \\ y & z \end{bmatrix} \quad \boldsymbol{J} = \begin{bmatrix} 2 & 3 & 4 \\ 1 & 4 & 2 \end{bmatrix}$$

1. If possible perform the following calculations:

(a) $\boldsymbol{A} + \boldsymbol{C}$ (b) $\boldsymbol{H} + \boldsymbol{B}$ (c) $\boldsymbol{J} + \boldsymbol{G}$ (d) $\boldsymbol{C} - \boldsymbol{H}$ (e) $\boldsymbol{A} - \boldsymbol{H}$ (f) $2\boldsymbol{C} + \boldsymbol{B}$ (g) $\frac{1}{2}\boldsymbol{G}$

2. If possible perform the following calculations:

(a) $\boldsymbol{AC}$ (b) $\boldsymbol{JG}$ (c) $\boldsymbol{GJ}$ (d) $\boldsymbol{EB}$ (e) $\boldsymbol{BE}$ (f) $\boldsymbol{GD}$ (g) $\frac{1}{2}\boldsymbol{HJ}$

3. Show that $(\boldsymbol{A} + \boldsymbol{B})^2 \neq \boldsymbol{A}^2 + 2\boldsymbol{AB} + \boldsymbol{B}^2$ but that $(\boldsymbol{A} + \boldsymbol{B})^2 = \boldsymbol{A}^2 + \boldsymbol{AB} + \boldsymbol{BA} + \boldsymbol{B}^2$.

4. (a) Find a matrix $\boldsymbol{I}$ for which the following equation holds:

$$\boldsymbol{I}\begin{bmatrix} a & b & c \\ d & e & f \end{bmatrix} = \begin{bmatrix} a & b & c \\ d & e & f \end{bmatrix}$$

(b) Find a matrix $\boldsymbol{I}$ for which the following equation holds:

$$\begin{bmatrix} a & b & c \\ d & e & f \end{bmatrix} = \boldsymbol{I}\begin{bmatrix} a & b & c \\ d & e & f \end{bmatrix}$$

5. A travel agent records the following bookings for types of holiday and destinations for the months of May and June.

Number of bookings in May

	Spain	Greece	Turkey	USA
Self catering	200	150	300	250
Hotel	100	80	150	100
Flight only	80	60	80	60

Number of bookings in June

	Spain	Greece	Turkey	USA
Self catering	230	165	370	300
Hotel	115	90	160	130
Flight only	95	70	95	70

The travel agent had projected that the bookings for June would have shown a 20% increase compared with those for May. Use matrices to calculate what the projected bookings for June were and calculate any shortfall in the actual bookings.

6. A small engineering company has two machines identified by M1 and M2 that are used to manufacture bolts. The company manufactures just three sizes of bolt. The following table shows the time in minutes that each of the machines requires to produce a bolt of a particular size.

Time in minutes to produce a bolt

	Machine	
	M1	M2
Large	3	2
Medium	2	1
Small	1.5	0.5

The cost of running these machines is given in the following table:

Cost per minute in pence

	Machine	
	M1	M2
Running cost	3	5

The company receives an order which is shown in the following table:

Order for bolts

	Large	Medium	Small
Quantity	200	300	100

The production manager decides to meet this order by splitting the manufacture of the bolts equally between the two machines. Using matrices calculate how long each machine will need to run to satisfy its part of the order. Calculate also the overall cost in running the machines.

7. Identify the elementary matrices that may be used to achieve the following transformations:

(a) $\begin{bmatrix} 4 & 6 & 7 \\ 2 & 1 & 5 \\ 2 & 2 & 3 \end{bmatrix}$ to $\begin{bmatrix} 0 & 2 & 1 \\ 2 & 1 & 5 \\ 2 & 2 & 3 \end{bmatrix}$

(b) $\begin{bmatrix} a & b \\ c & d \end{bmatrix}$ to $\begin{bmatrix} 2a & 2b \\ c & d \end{bmatrix}$

(c) $\begin{bmatrix} 1 & 0 & 0 \\ 0 & 1 & 0 \\ 0 & 0 & 1 \end{bmatrix}$ to $\begin{bmatrix} 1 & 0 & 0 \\ 0 & 0 & 1 \\ 0 & 1 & 0 \end{bmatrix}$

8. Find the inverses of the following matrices:

(a) $\begin{bmatrix} 4 & 2 \\ 5 & 3 \end{bmatrix}$ (b) $\begin{bmatrix} 5 & 2 \\ 3 & 4 \end{bmatrix}$ (c) $\begin{bmatrix} 1 & 1 & -1 \\ 2 & 1 & 1 \\ 3 & -2 & 1 \end{bmatrix}$ (d) $\begin{bmatrix} 2 & 1 & 1 \\ 3 & -2 & -2 \\ 4 & 2 & 3 \end{bmatrix}$

9. Use matrices to solve the following equations:

(a) $2x + y = 10$
$3x + 2y = 17$

(b) $3a + 4b = 11$
$a + 7b = 15$

(c) $2x - y + z = 12$
$4x + y - z = -6$
$2x - y + z = 18$

(d) $i + 2j - k = 20$
$3i + j + 2k = 30$
$2i + 3j + k = 40$

7.8 Further reading

J C Mason, *BASIC Matrix Methods* (Butterworths, 1984)
A text which contains a variety of matrix calculations implemented in the programming language BASIC.
Alan Jennings and J J McKeown, *Matrix Computation*, 2nd edn (Wiley, 1992)
A text which moves from the introductory to the intermediate level and which includes some computer program segments for the manipulation of matrices.
Stanley I Grossman, *Elementary Linear Algebra* (Wadsworth, 1987)
A more theoretical text than the previous two but one which starts at the introductory level.

8 Co-ordinate geometry

Objectives

When you have mastered this chapter you should be able to

- use co-ordinate geometry to specify the positions of and calculate the distance between points
- use co-ordinate geometry to describe the characteristics of lines and circles
- use matrices and homogeneous co-ordinates to calculate the effects of rotations and translations on points, lines and circles

In this chapter we will meet co-ordinate geometry and see how it can be used to describe points, lines and circles in two-dimensional space. We shall also consider the effects of transformations such as translations and rotations on these geometric objects. The contents of this chapter will provide a foundation for future work in the field of computer graphics.

8.1 The point

The fundamental element of computer graphics and co-ordinate geometry is the point. In one-dimensional space, i.e. on a line, the location of a point may be determined by its distance along that line from some fixed point called the **origin**. By convention distances to the right are given a positive value while those to the left are negative. The line along which points are located in one-dimensional space is called an **axis** and it is equivalent to the real number line. In Figure 8.1 we see that the origin O is located at position 0, point P is at location 5 and point Q is at location -3.5. The value that determines the

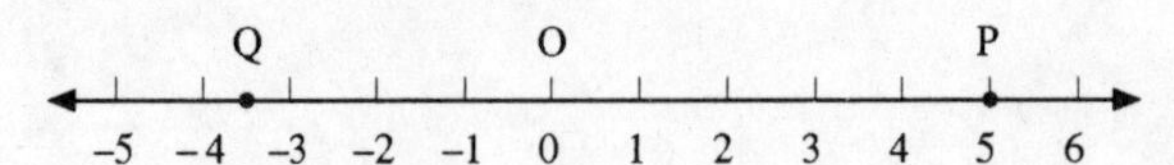

Figure 8.1
Locating points on an axis

location of a point relative to an origin is known as the point's **co-ordinate**. We will enclose co-ordinates in parentheses. So relative to the origin O the co-ordinate of P is (5) and the co-ordinate of Q is (–3.5).

Alternative origins

Consider the point O* which has a co-ordinate of (2) relative to the origin O. It is possible to specify the location of P relative to O* rather than O. If the co-ordinate of the point P relative to the origin O is represented by x_P and the co-ordinate of P relative to the origin O* is represented by x_P^* then from Figure 8.2 we see that the following is true:

$$x_P^* = x_P - 2$$

Substituting in numeric values we get

$$x_P^* = 5 - 2$$
$$x_P^* = 3$$

Hence the co-ordinate of P relative to O* is (3). Similarly for the point Q we have

$$x_Q^* = x_Q - 2$$

giving

$$x_Q^* = -3.5 - 2$$
$$x_Q^* = -5.5$$

So the co-ordinate of Q relative to O* is (–5.5).

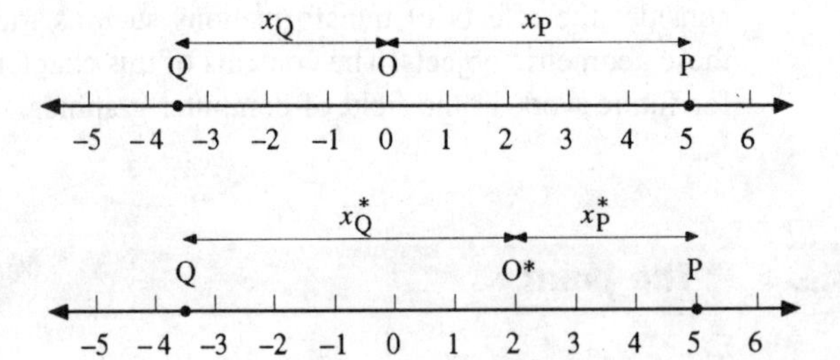

Figure 8.2
Moving the origin from O to O*

KEY POINT

In general, if the co-ordinate of some point P relative to the origin O located at (0) is x_P, then the co-ordinate x_P^* of the point relative to O* located at (T) is given by the equation

$$x_P^* = x_P - T$$

Worked examples

8.1 Indicate the locations of the following points on the real number line relative to an origin O located at position (0).

(a) Point P with co-ordinate (3.5). (b) Point Q with co-ordinate (–4).
(c) Point R with co-ordinate ($\sqrt{2.25}$).

Solution The locations of the points are shown in Figure 8.3.

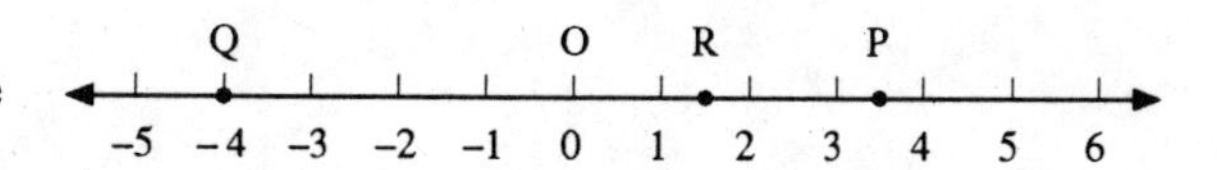

Figure 8.3
Solution to Worked example 8.1

8.2 Calculate the co-ordinates of the points P, Q and R in Worked example 8.1 relative to an origin O* which is located at (–2.5).

Solution Using the formula $x_P^* = x_P - T$ with T equal to –2.5, for the point P we have

$$x_P^* = 3.5 - (-2.5)$$

so

$$x_P^* = 6$$

For the point Q we have

$$x_Q^* = -4 - (-2.5)$$

so

$$x_Q^* = -1.5$$

For the point R we have

$$x_R^* = \sqrt{2.25} - (-2.5)$$

so

$$x_R^* = 4$$

Translating a point along the axis

Figure 8.4 shows a point located at P having a co-ordinate of (3), moved a distance of –7 to position P*, which has a co-ordinate of (–4). Such a movement is called a **translation**. In Figure 8.4 we can also see that the point located at Q having the co-ordinate (2) is translated through a distance of 4 to the new location Q* having the co-ordinate (6).

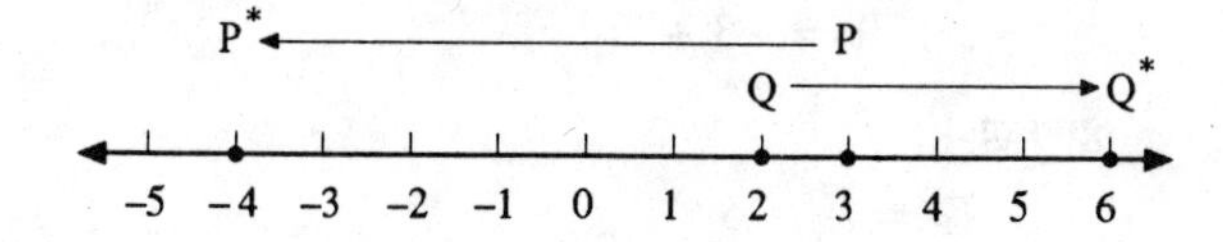

Figure 8.4
Translating a point

KEY POINT

In general if a point having a co-ordinate x_P is subjected to a translation of T, then x_P^*, the new co-ordinate of the point, is given by the equation

$$x_P^* = x_P + T$$

Worked examples

8.3 Calculate the new co-ordinates of the point P, having co-ordinate (3.5), and point Q, having co-ordinate (– 4.5), when they are subjected to a translation of – 3.5.

Solution In this case the value of T is – 3.5. Hence using the equation $x^* = x_P + T$ we get for P,

$$x_P^* = 3.5 + (-3.5)$$
$$x_P^* = 0$$

and for Q

$$x_Q^* = -4.5 + (-3.5)$$
$$x_Q^* = -8$$

8.4 The point P, having co-ordinate (2), and the point Q, having co-ordinate (–3), are subjected to the translations T_P and T_Q respectively, both points ending up at the same location. Derive the relationship between T_P and T_Q.

Solution For the point P we have

$$x_P^* = 2 + T_P$$

and for the point Q,

$$x_Q^* = -3 + T_Q$$

Now since the points end up at the same location we know that

$$x_P^* = x_Q^*$$

hence

$$2 + T_P = -3 + T_Q$$

giving

$$T_P = T_Q - 5$$

Distances between points

KEY POINT

To find the distance d between two points with co-ordinates x_p and x_q on the one-dimensional axis, the following formula may be used:

$$d = \sqrt{(x_p - x_q)^2}$$

The taking of the positive square root of the square of the difference ensures that all distances between points will be positive irrespective of the values of their co-ordinates.

Worked example

8.5 Calculate the distance d between the two points that have co-ordinates (−2.5) and (1.5).

Solution

$$d = \sqrt{(-2.5 - 1.5)^2} = \sqrt{(-4.0)^2} = \sqrt{16} = 4$$

So the distance between (−2.5) and (1.5) is 4.

Self-Assessment Questions 8.1

1. The point P has co-ordinate x_P relative to the origin O located at (0). What is the co-ordinate of P relative to the origin O* located at x_P?
2. The point P has co-ordinate x_P. What translation must P undergo for its co-ordinate to become $-x_P$?

Exercise 8.1

1. Relative to an origin O at location (0), the points P, Q and R have co-ordinates (−3.6), (4.5) and ($\sqrt{1.44}$) respectively. Calculate the co-ordinates of these points relative to the origin O*, given that O* has co-ordinate (2.4).
2. Calculate the translations that the points P and Q, located at (8) and (−2.5), respectively, must undergo in order to swap places.
3. Calculate the distances between the pairs of points having the following co-ordinates.

   (a) (4.5), (7.5)
   (b) (6), (− 6)
   (c) (− 3), (8)

8.2 Two dimensions

The introduction of a second dimension enables us to develop a system of co-ordinate geometry for describing and locating geometric shapes on a two-dimensional plane. The rest of this chapter deals with such a system.

The point in two-dimensional space

In order to locate a point in two-dimensional space, i.e. on a plane, two co-ordinate values and hence two axes are required. One method, though not the only one, is to choose two axes at right angles to each other and to locate the point by an ordered pair of numbers, usually written as (x, y). The first number indicates the distance of the point from the origin along the horizontal axis, called the ***x* co-ordinate** or **abscissa**, while the second number indicates the vertical distance of the point from the origin and is known as the ***y* co-ordinate** or **ordinate**. Such a system is called the **Cartesian co-ordinate system**, in honour of the French mathematician and philosopher René Descartes who laid the foundations of co-ordinate geometry.

Figure 8.5 shows the two axes x and y, intersecting at right angles at an origin O located at (0,0). The points P, Q, R and S have the co-ordinates (3, 3), (−3, 4), (−4, −3) and (5, −2) respectively.

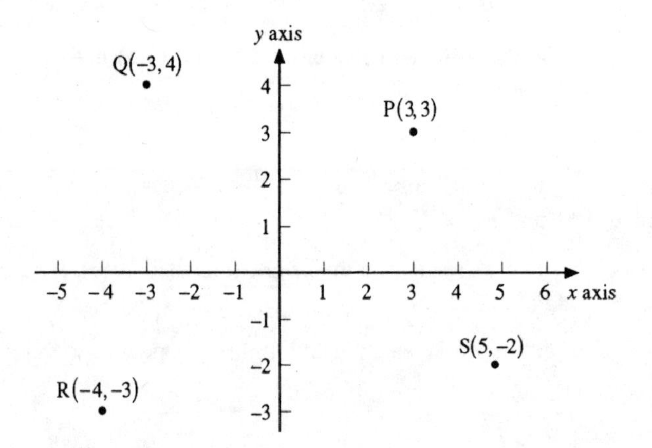

Figure 8.5
Locating points in two-dimensional space

Distance between points in two dimensions

You will be familiar with Pythagoras' theorem which tells us that in a right-angled triangle the square of the length of the hypotenuse (the side opposite to the right-angle) is equal to the sum of the squares of the lengths of the other two sides. This theorem may be used to calculate the distance between two points in two-dimensional space. In Figure 8.6 P and Q are two arbitrary points having co-ordinates (x_1, y_1) and (x_2, y_2) respectively. In order to calculate the

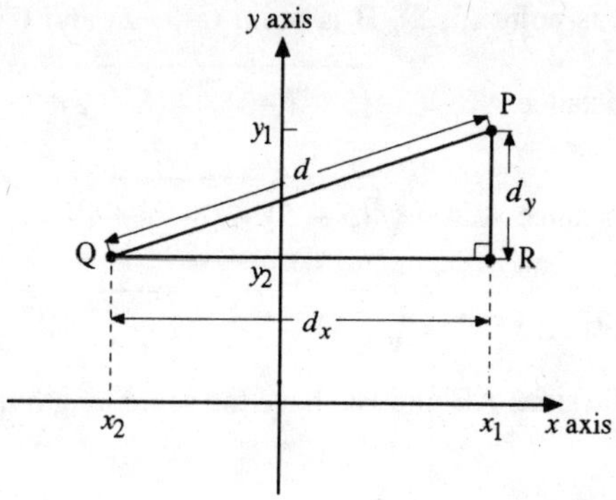

Figure 8.6
Distance between points in two-dimensional space

distance d between P and Q we first construct the right-angled triangle PRQ in which the side QR has a length of d_x and the side PR has a length of d_y.

Now applying Pythagoras' theorem we get

$$d = \sqrt{\left(d_x^2 + d_y^2\right)}$$

From the key point on page 169 we know that

$$d_x = \sqrt{(x_1 - x_2)^2} \quad \text{and that} \quad d_y = \sqrt{(y_1 - y_2)^2}$$

So we can now write the following.

KEY POINT

$$d = \sqrt{(x_1 - x_2)^2 + (y_1 - y_2)^2}$$

This formula enables us to calculate the distance between any two points in terms of their co-ordinates.

Worked examples

8.6 Calculate the distance between the two points that have co-ordinates (3, −1) and (−3, −4).

Solution Here we have $x_1 = 3$, $y_1 = -1$, $x_2 = -3$ and $y_2 = -4$. Substituting these values into the distance formula we get

$$d = \sqrt{(3 - (-3))^2 + (-1 - (-4))^2} = \sqrt{(36 + 9)} = \sqrt{45} = 3\sqrt{5}$$

Hence the distance between the points (3, −1) and (−3, −4) is $3\sqrt{5}$ which is approximately 6.708.

8.7 Show that the triangle ABC whose vertices are located at the points (5, 8), (−3, 2) and (5, −4) is isosceles, i.e. two sides of the triangle are equal in length.

Solution Now A is point (5, 8), B is point (−3, 2) and C is point (5, −4). Hence:

$$\text{distance AB} = \sqrt{(5-(-3))^2+(8-2)^2} = \sqrt{(64+36)} = 10$$

$$\text{distance AC} = \sqrt{(5-5)^2+(8-(-4))^2} = \sqrt{(0+144)} = 12$$

$$\text{distance BC} = \sqrt{(-3-5)^2+(2-(-4))^2} = \sqrt{(64+36)} = 10$$

Since the sides AB and BC have the same length the triangle ABC is isosceles.

Self-Assessment Question 8.2

Write down the expression giving the distance between the point P located at (x_P, y_P) and the origin.

Exercise 8.2

1. On a sketch graph indicate the locations of the points P, Q, R and S which have the following co-ordinates: (2, −3), (−5, 4), (−3, −3) and (3, 5) respectively.
2. Calculate the distance between the two points which have the co-ordinates (10, 7) and (−2, 2).
3. Show that the triangle with vertices A, B and C located at the points (−7, 5), (3, −1) and (1, 7) is right angled.

8.3 Transformations in two dimensions

In this section we will see how to describe translations and rotations in two-dimensional space. In particular we shall see how matrices provide a compact notation for describing and combining such transformations.

Translation of the origin

A point P having co-ordinates (x_P, y_P) relative to an origin O with co-ordinates (0,0) and the axes x and y is shown in Figure 8.7.

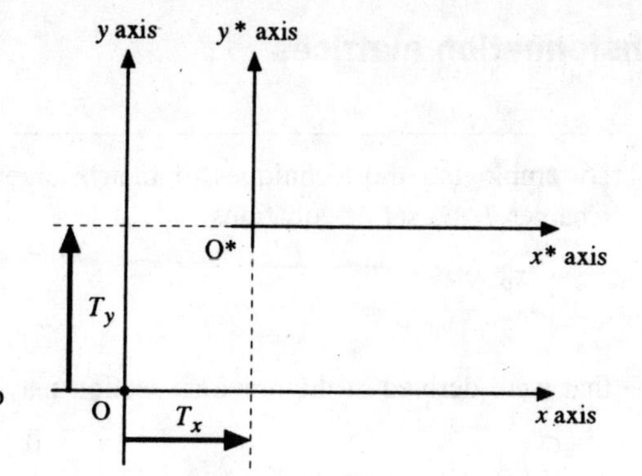

Figure 8.7
Translating the origin in two dimensions

KEY POINT

If the origin O is translated a distance T_y in the vertical direction and a distance T_x in the horizontal direction to give the new origin O*, then the co-ordinates of P, (x_P^*, y_P^*) relative to O* and the axes x^* and y^*, are given by

$$x_P^* = x_P - T_x$$
$$y_P^* = y_P - T_y$$

Worked example

8.8 A point has co-ordinates (5, −7) relative to an origin O located at (0,0). If the origin is translated to O*, located at (4, −3), calculate the co-ordinates of the point in relation to this new origin.

Solution In this case $T_x = 4$ and $T_y = -3$, hence the new co-ordinates $(x^*,\ y^*)$ are given by

$$x_P^* = x_P - T_x$$
$$y_P^* = y_P - T_y$$

Hence:

$$x_P^* = 5 - 4$$
$$x_P^* = 1$$

and

$$y_P^* = -7 - (-3)$$
$$y_P^* = -4$$

So the co-ordinates of P relative to O* are (1, −4).

Transformation matrices

KEY POINT

By employing the techniques of matrix algebra that we met in Chapter 7 the set of equations

$$x_P^* = x_P - T_x$$
$$y_P^* = y_P - T_y$$

that were derived in the previous section may be written as

$$[x^* \quad y^* \quad 1] = [x \quad y \quad 1] \begin{bmatrix} 1 & 0 & 0 \\ 0 & 1 & 0 \\ -T_x & -T_y & 1 \end{bmatrix}$$

The triple $[x \quad y \quad 1]$ is known as the **homogeneous co-ordinate form** of the co-ordinate pair (x, y), while the 3×3 matrix is known as the **transformation matrix**.

Matrix notation not only provides a compact way of describing the transformation, it also suggests the data structures and routines that may be used to represent such a transformation in a computer program.

Worked example

8.9 A point has co-ordinates $(3, -5)$ relative to an origin O located at (0,0). The origin is translated to O* located at $(-5, 6)$. Use homogeneous co-ordinates and a transformation matrix to calculate the co-ordinates of the point in relation to this new origin.

Solution Using

$$[x^* \quad y^* \quad 1] = [x \quad y \quad 1] \begin{bmatrix} 1 & 0 & 0 \\ 0 & 1 & 0 \\ -T_x & -T_y & 1 \end{bmatrix}$$

with $T_x = -5$, $T_y = 6$, $x = 3$ and $y = -5$ we get

$$[x^* \quad y^* \quad 1] = [3 \quad -5 \quad 1] \begin{bmatrix} 1 & 0 & 0 \\ 0 & 1 & 0 \\ 5 & -6 & 1 \end{bmatrix}$$

$$[x^* \quad y^* \quad 1] = [3 + 0 + 5 \quad 0 + (-5) + (-6) \quad 0 + 0 + 1]$$

$$[x^* \quad y^* \quad 1] = [8 \quad -11 \quad 1]$$

Hence $(x^*, y^*) = (8, -11)$.

Translating a point

Transformation matrices and homogeneous co-ordinates may also be used to determine the co-ordinates of a point P relative to the origin O at (0, 0) after the point has been translated across the two-dimensional plane. In Figure 8.8 the point P with co-ordinates (x,y) has been translated to the location P* with co-ordinates (x^*, y^*) by a horizontal translation of T_x and a vertical translation of T_y.

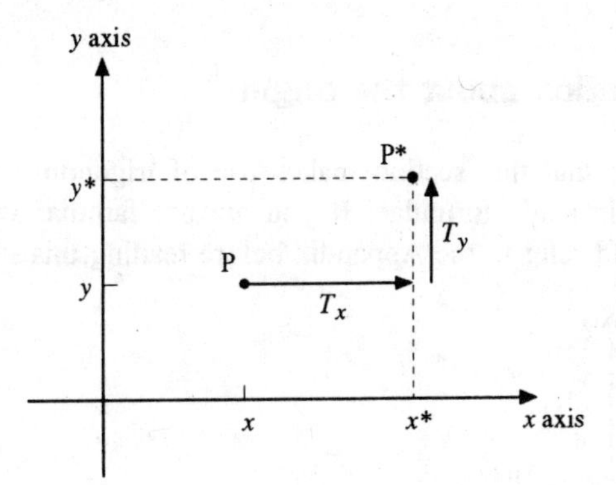

Figure 8.8
Translating a point in two dimensions

KEY POINT

From Figure 8.8 we see that

$$x^* = x + T_x$$
$$y^* = y + T_y$$

This pair of equations may be written as

$$[x^* \quad y^* \quad 1] = [x \quad y \quad 1]\begin{bmatrix} 1 & 0 & 0 \\ 0 & 1 & 0 \\ T_x & T_y & 1 \end{bmatrix}$$

Worked example

8.10 A point P has co-ordinates (5, −8). Determine the co-ordinates of this point if it is subjected to a horizontal translation of −8 and a vertical translation of 4.

Solution Using

$$[x^* \quad y^* \quad 1] = [x \quad y \quad 1]\begin{bmatrix} 1 & 0 & 0 \\ 0 & 1 & 0 \\ T_x & T_y & 1 \end{bmatrix}$$

with $T_x = -8$, $T_y = 4$, $x = 5$ and $y = -8$ we get

$$\begin{aligned}[x^* \quad y^* \quad 1] &= [5 \quad -8 \quad 1] \begin{bmatrix} 1 & 0 & 0 \\ 0 & 1 & 0 \\ -8 & 4 & 1 \end{bmatrix} \\ &= [5+0+(-8) \quad 0+(-8)+4 \quad 0+0+1] \\ &= [-3 \quad -4 \quad 1]\end{aligned}$$

Hence $(x^*, y^*) = (-3, -4)$.

Rotation about the origin

(Note that this section makes use of trigonometric identities, in particular double angle formulae. If you are not familiar with these topics then you should refer to the Appendix before reading this section.)

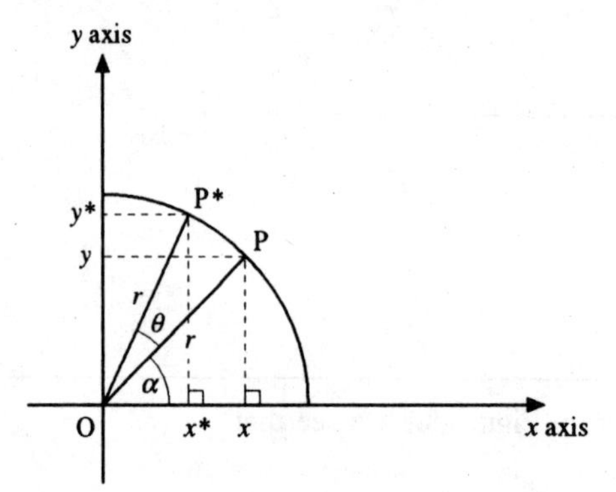

Figure 8.9
Rotating a point about the origin

In Figure 8.9 the point P has been transformed to P* by an anti-clockwise rotation through an angle of θ degrees about the origin. By convention a positive angle represents an anti-clockwise rotation while a negative angle represents a clockwise rotation. From Figure 8.9 we see that if r is the length of the line joining the point P* to the origin and if this line makes an angle of $\alpha + \theta$ with the x-axis then

$$x^* = r\cos(\alpha + \theta) \quad \text{and} \quad y^* = r\sin(\alpha + \theta)$$

Using the fact that $\cos(\alpha + \theta) = \cos\alpha\cos\theta - \sin\alpha\sin\theta$ and $\sin(\alpha + \theta) = \sin\alpha\cos\theta + \cos\alpha\sin\theta$ we can write

$$x^* = r(\cos\alpha\cos\theta - \sin\alpha\sin\theta)$$

and

$$y^* = r(\sin\alpha\cos\theta + \cos\alpha\sin\theta)$$

so

$$x^* = r\cos\alpha\cos\theta - r\sin\alpha\sin\theta$$
$$y^* = r\sin\alpha\cos\theta + r\cos\alpha\sin\theta$$

From Figure 8.9 we see that the length of the line joining the point P to the origin is also r and that this line makes an angle of α with the x axis. Hence we have

$$x = r\cos\alpha \quad \text{and} \quad y = r\sin\alpha$$

So we can now write

$$x^* = x\cos\theta - y\sin\theta$$
$$y^* = y\cos\theta + x\sin\theta$$

which is the same as

$$x^* = x\cos\theta - y\sin\theta$$
$$y^* = x\sin\theta + y\cos\theta$$

Using matrices the above pair of equations may be written as

$$[x^* \quad y^*] = [x \quad y]\begin{bmatrix} \cos\theta & \sin\theta \\ -\sin\theta & \cos\theta \end{bmatrix}$$

KEY POINT

Using homogeneous co-ordinates this equation becomes

$$[x^* \quad y^* \quad 1] = [x \quad y \quad 1]\begin{bmatrix} \cos\theta & \sin\theta & 0 \\ -\sin\theta & \cos\theta & 0 \\ 0 & 0 & 1 \end{bmatrix}$$

Worked example

8.11 A point P has co-ordinates (8, –6). Determine the co-ordinates of this point after it has been subjected to a rotation of 60° about the origin.

Solution Using

$$[x^* \quad y^* \quad 1] = [x \quad y \quad 1]\begin{bmatrix} \cos\theta & \sin\theta & 0 \\ -\sin\theta & \cos\theta & 0 \\ 0 & 0 & 1 \end{bmatrix}$$

we get

$$[x^* \quad y^* \quad 1] = [8 \quad -6 \quad 1]\begin{bmatrix} \cos 60^\circ & \sin 60^\circ & 0 \\ -\sin 60^\circ & \cos 60^\circ & 0 \\ 0 & 0 & 1 \end{bmatrix}$$

$$= [8 \quad -6 \quad 1]\begin{bmatrix} 1/2 & \sqrt{3}/2 & 0 \\ -\sqrt{3}/2 & 1/2 & 0 \\ 0 & 0 & 1 \end{bmatrix}$$

$$= \left[4 + 3\sqrt{3} \quad 4\sqrt{3} - 3 \quad 1\right]$$

giving

$$(x^*, y^*) = \left(4 + 3\sqrt{3},\ 4\sqrt{3} - 3\right)$$

Composite transformations

In this section we will see how transformation matrices may be combined in order to describe composite transformations.

To illustrate this process consider the problem of determining the location of a straight line after it has been rotated about some arbitrary point in the plane. Figure 8.10 shows a straight line which is completely specified by its end-points located at (6,6) and $(-2, 4)$. We can determine the location of this line after it has been rotated through an angle of 30° about the point P located at co-ordinates (2,8) by applying this transformation to the two points (6,6) and $(-2, 4)$ that specify the line. The two new points produced by this transformation will specify the position of the transformed line.

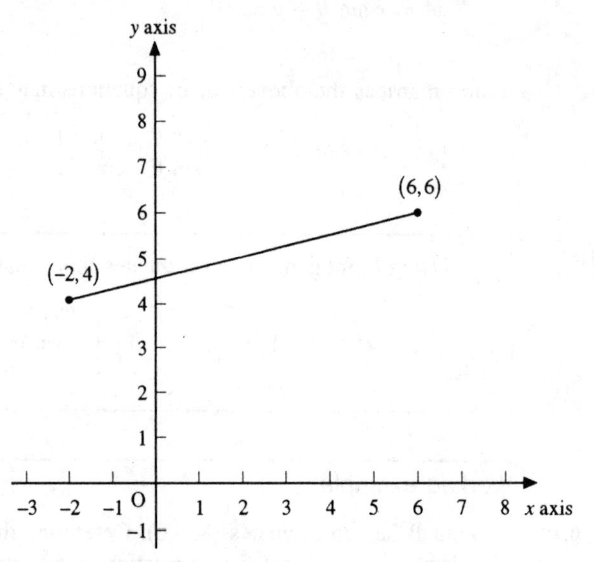

Figure 8.10
A line specified by its end-points

The process of rotation about a point which is not the origin may be achieved in three steps. The first step is to translate the origin to the point of rotation, the second is to perform the rotation about the new origin and finally the third step is to translate the origin to back to its initial position.

Step 1 Moving the origin to the point of rotation, in this case (2,8)

The effect on a line of translating the origin can be determined by applying the appropriate transformation matrix to any two distinct points that lie on the line. In this case the transformation matrix $\boldsymbol{T}$ is

$$\boldsymbol{T} = \begin{bmatrix} 1 & 0 & 0 \\ 0 & 1 & 0 \\ -2 & -8 & 1 \end{bmatrix}$$

The results of applying this matrix to the two points (6,6) and $(-2, 4)$ are shown in Figure 8.11. The calculation of the transformation is as follows. For $(-2, 4)$,

$$[x^* \quad y^* \quad 1] = [-2 \quad 4 \quad 1] \begin{bmatrix} 1 & 0 & 0 \\ 0 & 1 & 0 \\ -2 & -8 & 1 \end{bmatrix} = [-4 \quad -4 \quad 1]$$

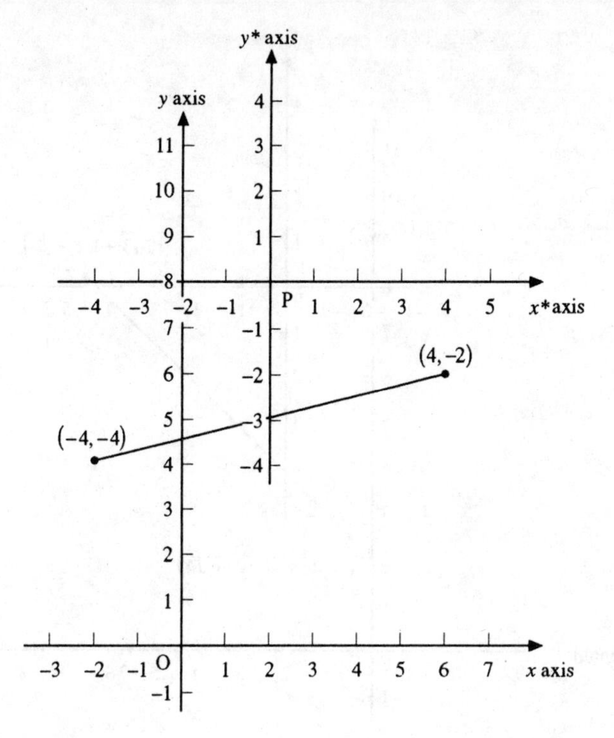

Figure 8.11
A line specified relative to a translated origin

so (–2, 4) becomes (–4, –4). For (6,6),

$$[x^* \quad y^* \quad 1] = [6 \quad 6 \quad 1]\begin{bmatrix} 1 & 0 & 0 \\ 0 & 1 & 0 \\ -2 & -8 & 1 \end{bmatrix} = [4 \quad -2 \quad 1]$$

so (6,6) becomes (4, –2).

Step 2 Rotating the line about the new origin P

The effect of rotating the line about the origin P can be determined by applying the following transformation matrix $\boldsymbol{R}$ to the two points that lie on the line.

$$\boldsymbol{R} = \begin{bmatrix} \cos\theta & \sin\theta & 0 \\ -\sin\theta & \cos\theta & 0 \\ 0 & 0 & 1 \end{bmatrix}$$

In this particular case the rotation is anti-clockwise through an angle of 30° so the transformation matrix becomes

$$\begin{bmatrix} \sqrt{3}/2 & 1/2 & 0 \\ -1/2 & \sqrt{3}/2 & 0 \\ 0 & 0 & 1 \end{bmatrix}$$

The results of applying this transformation matrix to the two points, which are now (–4, –4) and (4, –2), are shown in Figure 8.12. The calculation of the

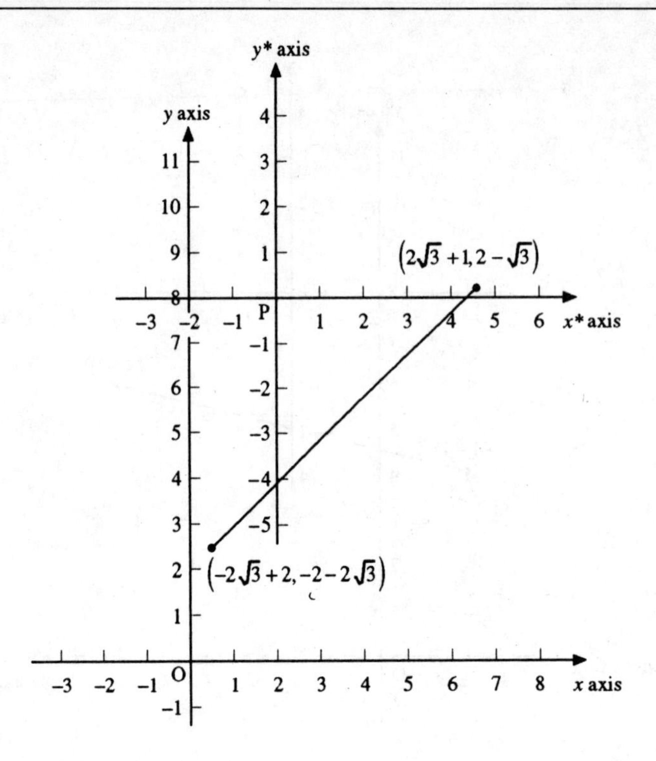

Figure 8.12
A line having been rotated about a new origin

transformation is as follows. For $(-4, -4)$,

$$[x^* \quad y^* \quad 1] = [-4 \quad -4 \quad 1]\begin{bmatrix} \sqrt{3}/2 & 1/2 & 0 \\ -1/2 & \sqrt{3}/2 & 0 \\ 0 & 0 & 1 \end{bmatrix}$$

$$= [-2\sqrt{3}+2 \quad -2-2\sqrt{3} \quad 1]$$

so $(-4, -4)$ becomes $(-2\sqrt{3}+2, -2-2\sqrt{3})$. For $(4, -2)$,

$$[x^* \quad y^* \quad 1] = [4 \quad -2 \quad 1]\begin{bmatrix} \sqrt{3}/2 & 1/2 & 0 \\ -1/2 & \sqrt{3}/2 & 0 \\ 0 & 0 & 1 \end{bmatrix}$$

$$= [2\sqrt{3}+1 \quad 2-\sqrt{3} \quad 1]$$

so $(4, -2)$ becomes $(2\sqrt{3}+1, 2-\sqrt{3})$.

Step 3 Move the origin back to (0,0)

This is the inverse of the transformation that took the origin to (2,8). This transformation can be achieved by applying the inverse of the original transformation matrix to the end-points of the line. This inverse matrix is

$$\begin{bmatrix} 1 & 0 & 0 \\ 0 & 1 & 0 \\ 2 & 8 & 1 \end{bmatrix}$$

since

$$\begin{bmatrix} 1 & 0 & 0 \\ 0 & 1 & 0 \\ -2 & -8 & 1 \end{bmatrix} \begin{bmatrix} 1 & 0 & 0 \\ 0 & 1 & 0 \\ 2 & 8 & 1 \end{bmatrix} = \begin{bmatrix} 1 & 0 & 0 \\ 0 & 1 & 0 \\ 0 & 0 & 1 \end{bmatrix}$$

The results of applying this matrix to the two points, which are $(-2\sqrt{3}+2, -2-2\sqrt{3})$ and $(2\sqrt{3}+1, 2-\sqrt{3})$, are shown in Figure 8.13.

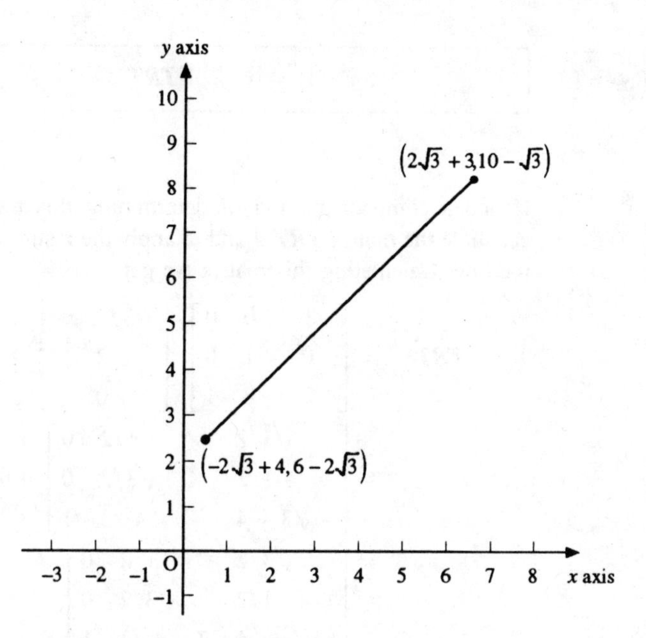

Figure 8.13
Rotated line shown relative to origin at (0, 0)

The calculation of the transformation is as follows.
For $(-2\sqrt{3}+2, -2-2\sqrt{3})$,

$$\begin{aligned} [x^* \quad y^* \quad 1] &= [-2\sqrt{3}+2 \quad -2-2\sqrt{3} \quad 1] \begin{bmatrix} 1 & 0 & 0 \\ 0 & 1 & 0 \\ 2 & 8 & 1 \end{bmatrix} \\ &= [-2\sqrt{3}+4 \quad 6-2\sqrt{3} \quad 1] \end{aligned}$$

so $(-2\sqrt{3}+2, -2-2\sqrt{3})$ becomes $(-2\sqrt{3}+4, 6-2\sqrt{3})$.
For $(2\sqrt{3}+1, 2-\sqrt{3})$,

$$\begin{aligned} [x^* \quad y^* \quad 1] &= [2\sqrt{3}+1 \quad 2-\sqrt{3} \quad 1] \begin{bmatrix} 1 & 0 & 0 \\ 0 & 1 & 0 \\ 2 & 8 & 1 \end{bmatrix} \\ &= [2\sqrt{3}+3 \quad 10-\sqrt{3} \quad 1] \end{aligned}$$

so $(2\sqrt{3}+1, 2-\sqrt{3})$ becomes $(2\sqrt{3}+3, 10-\sqrt{3})$. The overall result of this composite transformation is that the points $(-2, 4)$ and (6,6) become $(-2\sqrt{3}+4, 6-2\sqrt{3})$ and $(2\sqrt{3}+3, 10-\sqrt{3})$ respectively. These two new points specify the position of the transformed line.

The rotation of the line about the point (2,8) was achieved by first applying a transformation matrix $\boldsymbol{T}$ which moved the origin to (2,8), and then by applying the matrix $\boldsymbol{R}$ which performed the rotation through 30° and finally applying the matrix $\boldsymbol{T}^{-1}$ which returned the origin to (0,0). The result of applying these transformations to a point (x,y) can be determined by the following calculation.

KEY POINT

$$[x^*,\ y^*,\ 1] = [x,\ y,\ 1]\boldsymbol{TRT}^{-1}$$

Hence a compact method of determining this transformation would be to calculate the matrix $\boldsymbol{TRT}^{-1}$ and to apply the result to the two points that lie on the line. Calculating this matrix we get

$$\boldsymbol{TRT}^{-1} = \begin{bmatrix} 1 & 0 & 0 \\ 0 & 1 & 0 \\ -2 & -8 & 1 \end{bmatrix} \begin{bmatrix} \sqrt{3}/2 & 1/2 & 0 \\ -1/2 & \sqrt{3}/2 & 0 \\ 0 & 0 & 1 \end{bmatrix} \begin{bmatrix} 1 & 0 & 0 \\ 0 & 1 & 0 \\ 2 & 8 & 1 \end{bmatrix}$$

$$= \begin{bmatrix} \sqrt{3}/2 & 1/2 & 0 \\ -1/2 & \sqrt{3}/2 & 0 \\ -\sqrt{3}+4 & -1-4\sqrt{3} & 1 \end{bmatrix} \begin{bmatrix} 1 & 0 & 0 \\ 0 & 1 & 0 \\ 2 & 8 & 1 \end{bmatrix}$$

$$= \begin{bmatrix} \sqrt{3}/2 & 1/2 & 0 \\ -1/2 & \sqrt{3}/2 & 0 \\ -\sqrt{3}+6 & 7-4\sqrt{3} & 1 \end{bmatrix}$$

This is the transformation matrix which will perform a rotation of 30° about the point (2,8). We can check that this matrix performs as expected by applying it to one of the points of the above line. Applying the matrix to the point $(-2,4)$ we get

$$[x^* \quad y^* \quad 1] = [-2 \quad 4 \quad 1] \begin{bmatrix} \sqrt{3}/2 & 1/2 & 0 \\ -1/2 & \sqrt{3}/2 & 0 \\ -\sqrt{3}+6 & 7-4\sqrt{3} & 1 \end{bmatrix}$$

$$= [-2\sqrt{3}+4 \quad 6-2\sqrt{3} \quad 1]$$

So $(-2,4)$ becomes $(-2\sqrt{3}+4, 6-2\sqrt{3})$ which is consistent with the result obtained above.

Self-Assessment Questions 8.3

1. Write down the transformation matrix that translates the point P located at (x_P, y_P) to the origin.
2. What matrix represents a rotation of 360° about the origin?
3. Is it generally true that applying a given translation followed by a rotation to a point has the same effect as applying the rotation followed by the translation? Justify your answer.

Exercise 8.3

1. Points P and Q are located at co-ordinates $(-3, 9)$ and $(2, -4)$ relative to an origin O at (0,0). Calculate the co-ordinates of P and Q relative to the origin O* located at $(-6, 2)$.
2. Show that the distance between any two points P and Q is unchanged when the origin moves from O to some arbitrary location O*.
3. The position of a triangle in relation to the origin O at (0,0) is defined by the co-ordinates of its three vertices. They are (2,5), $(-3, 7)$ and $(5, -8)$. By using a transformation matrix determine the co-ordinates of these vertices in relation to an origin O* located at $(5, -8)$.
4. A triangle with vertices A, B and C having co-ordinates (4, 6), $(-2, 8)$ and $(-6, 2)$ is subjected to a transformation consisting of a horizontal translation of -2 and a vertical translation of 4. Using a transformation matrix calculate the new co-ordinates of the vertices A, B and C and sketch the location of the transformed triangle.
5. A triangle with vertices A, B and C having co-ordinates (4,6), $(-2, 8)$ and $(-6, 2)$ is subjected to a rotation of 45° about the origin. By calculating the new co-ordinates of the vertices A, B and C sketch the location of the transformed triangle (note that $\cos 45° = \sqrt{2}/2$ and $\sin 45° = \sqrt{2}/2$).
6. Calculate the distance between the points $(-2, -4)$ and (2,6) and show that the distance is unchanged when the points are rotated through an angle of 60° about the origin.
7. Apply the transformation matrix $\boldsymbol{TRT}^{-1}$ which has been calculated in this section to the point (6,6) to show that its transformation is also consistent with the earlier calculation.
8. The result of applying a rotation to a polygon can be determined by calculating the effects of the rotation on the vertices of the polygon. As an example consider the triangle with vertices located at $(-2, 3)$, (5,10) and (8,1). Determine the result of rotating this triangle through 90° about the point (4,8) by applying the appropriate transformation matrix to its vertices. Produce a sketch showing the triangle before and after this rotation.

8.4 The straight line

So far we have specified straight lines by giving the co-ordinates of two points that lie on a particular line. In this section we will develop more general ways of specifying a straight line.

The equation of a straight line

Figure 8.14 shows a non-vertical straight line on which are located two distinct points P_1 and P_2 having co-ordinates (x_1,y_1) and (x_2,y_2) respectively, and a general point P with co-ordinates (x,y).

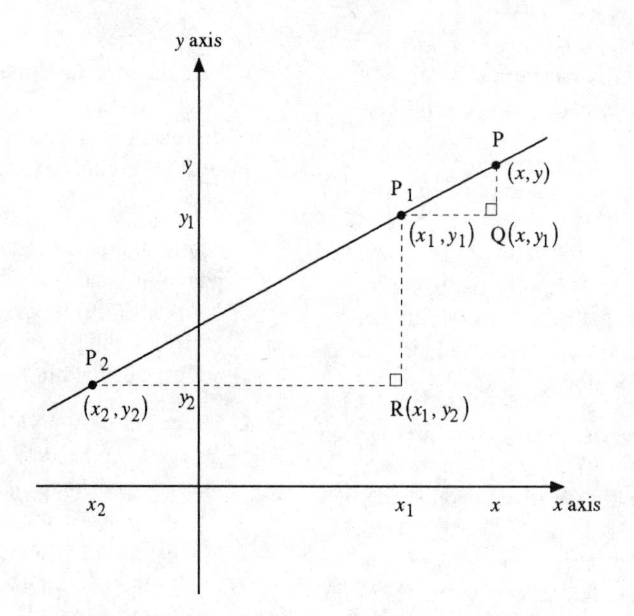

Figure 8.14
A line whose equation is $(y-y_1)(x_1-x_2) = (y_1-y_2)(x-x_1)$

The lines P_1Q and P_2R are parallel to each other, and to the x-axis. The two angles PP_1Q and P_1P_2R must be equal to each other and to the angle that the line makes with the x-axis. Now since the triangles PP_1Q and P_1P_2R are right angled we can state that

$$\tan(PP_1Q) = \tan(P_1P_2R)$$

so

$$\frac{y-y_1}{x-x_1} = \frac{y_1-y_2}{x_1-x_2}$$

This equation is valid as long as x_1 and x_2 are not equal, which is the case here since the points P_1 and P_2 are distinct and the straight line is not vertical (the case when they are equal is dealt with below).

KEY POINT

This equation may be written as

$$(y - y_1)(x_1 - x_2) = (y_1 - y_2)(x - x_1)$$

This is the general equation of a straight line, specified by the co-ordinates of two points (x_1, y_1) and (x_2, y_2) that lie on the line. This form of the straight line equation is often called the **two-point form**.

Worked example

8.12 Find the equation of the straight line that passes through the points $(-2, 4)$ and $(-5, -8)$.

Solution Substituting the given co-ordinates into the equation

$$(y - y_1)(x_1 - x_2) = (y_1 - y_2)(x - x_1)$$

gives

$$(y - 4)(-2 - (-5)) = (4 - (-8))(x - (-2))$$

so

$$3(y - 4) = 12(x + 2)$$

giving

$$3y - 12 = 12x + 24$$

which can be written as

$$3y - 12x - 36 = 0$$

The equation of a vertical line

Figure 8.15 shows a vertical straight line which cuts the x axis at $(a,0)$. The point P with co-ordinates (x,y) lies on the line. Now since the line is vertical all

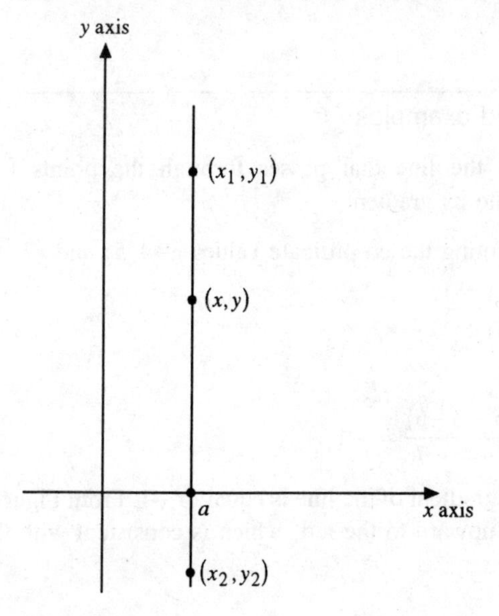

Figure 8.15
A line whose equation is $x = a$

points on the line must have the same x co-ordinate so we can write

$$x_1 = x_2 = x = a \quad \text{or just simply} \quad x = a$$

This is the equation of a vertical straight line which cuts the x axis at $(a,0)$.

The gradient-point form

The equation

$$(y - y_1)(x_1 - x_2) = (y_1 - y_2)(x - x_1)$$

from the key point on page 185 may be written in the form

$$(y - y_1) = \frac{y_1 - y_2}{x_1 - x_2}(x - x_1)$$

From the previous section you may recall that the ratio

$$\frac{y_1 - y_2}{x_1 - x_2}$$

is a constant for any given line, and is equal to the tangent of the angle between the line and the x axis.

KEY POINT

This ratio is known as the **gradient** of the line and is often replaced by the single letter m to give the following form:

$$y - y_1 = m(x - x_1)$$

This is sometimes referred to as the **gradient-point form** of the staight line equation.

Worked examples

8.13 Sketch the line that passes through the points $(-4, 5)$ and $(7, -6)$, and calculate its gradient.

Solution Substituting the co-ordinate values $(-4, 5)$ and $(7, -6)$ into

$$\frac{y_1 - y_2}{x_1 - x_2}$$

we get

$$\frac{5 - (-6)}{-4 - 7} = -1$$

So the gradient of the line is equal to -1. From Figure 8.16 we see that the line slopes upward to the left, which is consistent with the line having a negative gradient.

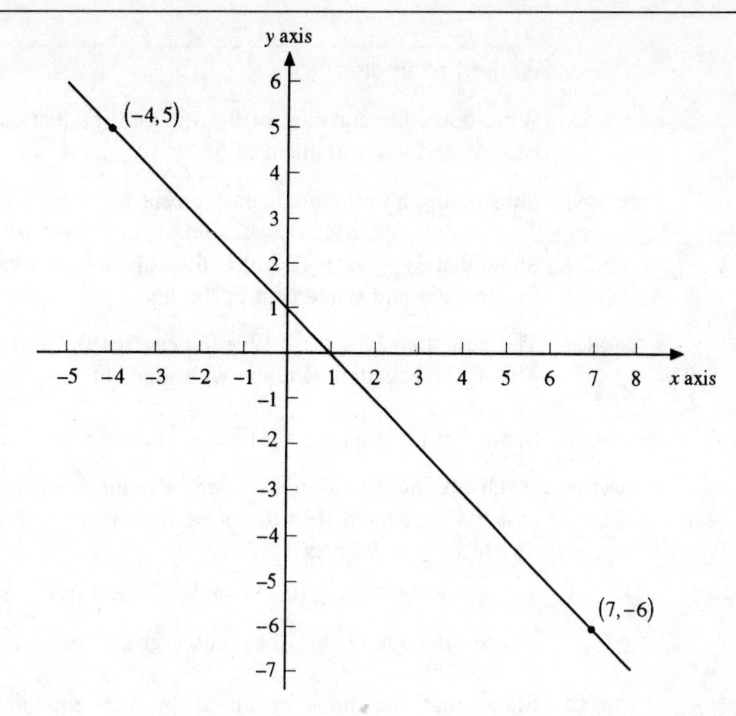

Figure 8.16
A line with a gradient of −1

8.14 Find the equation of the straight line with gradient −2 that passes through the point (3, −2).

Solution Using the gradient-point form $y - y_1 = m(x - x_1)$ we get

$$y - (-2) = -2(x - 3) \quad \text{which gives} \quad y + 2 = -2x + 6$$

which is

$$y = -2x + 4$$

The gradient-intercept form

Consider the straight line with a gradient of m that passes through the point (0, c). That is to say the line cuts the y axis at a distance c from the origin. What is the equation of this line?

KEY POINT

We can use the gradient-point form to write

$$y - c = m(x - 0)$$

giving

$$y = mx + c$$

This is known as the **gradient-intercept** form of the straight line, since the line 'intercepts' the y axis at the point (0, c). This form of the straight line equation is just a special case of the gradient-point form.

Worked examples

8.15 Write down the equation of the straight line that cuts the y axis at the point $(0, -6)$ and has a gradient of 3.

Solution Substituting into the gradient-intercept form gives $y = 3x - 6$.

8.16 Show that $3y - 9x + 12 = 0$ is the equation of a straight line and determine the gradient and y intercept of the line.

Solution The equation $3y - 9x + 12 = 0$ becomes $3y = 9x - 12$ giving $y = 3x - 4$. This is the equation of a line with gradient 3 and y intercept $(0, -4)$.

8.17 Show that the point $(1, 0.5)$ lies on the line $2y + 4x - 5 = 0$.

Solution If substituting the co-ordinates of the point into the equation of the line results in a true statement then the point lies on the line. Substituting $(1, 0.5)$ into $2y + 4x - 5 = 0$ gives

$$2y + 4x - 5 = 2(0.5) + 4(1) - 5 = 1 + 4 - 5 = 0$$

Hence the point $(1, 0.5)$ lies on the given line.

8.18 Show that the lines specified by the equations $2y - 3x + 6 = 0$ and $4y = 6x + 3$ are parallel.

Solution Re-writing each equation in the gradient-intercept form we get

$$\begin{aligned} 2y - 3x + 6 &= 0 \\ 2y &= 3x - 6 \\ y &= \tfrac{3}{2}x - 3 \end{aligned} \qquad \text{and} \qquad \begin{aligned} 4y &= 6x + 3 \\ y &= \tfrac{6}{4}x + \tfrac{3}{4} \\ y &= \tfrac{3}{2}x + \tfrac{3}{4} \end{aligned}$$

The gradients of both lines are equal to 3/2 and so the lines are parallel.

8.19 Find the equation of the line that passes through the point $(-2, -3)$ and is parallel to the line $3x - 7y + 4 = 0$.

Solution Re-writing the equation $3x - 7y + 4 = 0$ in gradient-intercept form gives $-7y = -3x - 4$. Dividing both sides by -7 we get $y = 3/7x + 4/7$. This gives a gradient of 3/7. So using the gradient-point form with point $(-2, -3)$ and gradient 3/7 we get:

$$\begin{aligned} y - y_1 &= m(x - x_1) \\ y - (-3) &= \tfrac{3}{7}(x - (-2)) \\ y + 3 &= \tfrac{3}{7}(x + 2) \\ 7y + 21 &= 3x + 6 \end{aligned}$$

giving

$$7y - 3x + 15 = 0$$

Self-Assessment Questions 8.4

1. What is the equation of a horizontal straight line that cuts the y axis at the point $(0,b)$?
2. What is the relationship between the gradients of two lines that are parallel?
3. What is the gradient of the straight line that cuts the x axis at $(a,0)$ and the y axis at $(0,a)$? Write down the equation of this line.

Exercise 8.4

1. Determine the equations of the straight lines that pass through the following pairs of points.
(a) (2,1) and (3,4) (b) (2,2) and (−3, −13)
2. Derive the equation of the straight line that passes through the origin and through the point (2,8).
3. Calculate the location at which the line that passes through (2,7) and (−4, −5) cuts the the y axis.
4. Calculate the equation of the line that is parallel to the line $3y + 9x - 6 = 0$ and cuts the y axis at (0,5).

8.5 The circle

By definition a circle consists of the set of points all of which are the same distance, called the **radius**, from some fixed point called the **centre**.

The equation of a circle

Figure 8.17 shows a circle with a radius of length r and a centre located at the point (a,b). The point (x,y) is some general point on the circumference of the

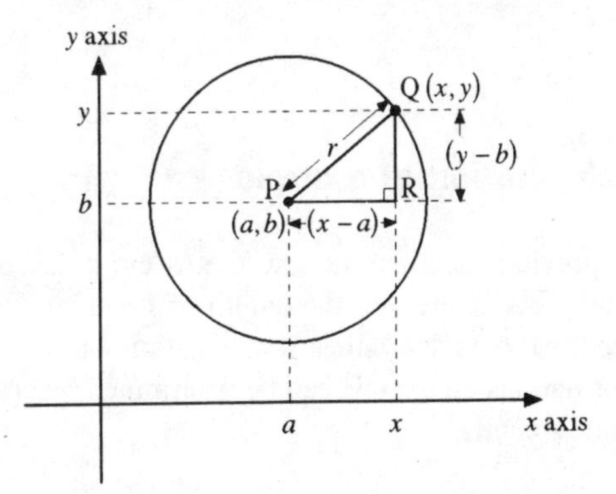

Figure 8.17
A circle whose equation is $(x-a)^2 + (y-b)^2 = r^2$

circle. Applying Pythagoras' theorem to the triangle PQR we get the following equation:

KEY POINT

$$(x-a)^2 + (y-b)^2 = r^2$$

This is the equation of a circle which has a radius of r and a centre at (a,b).

Worked examples

8.20 Find the equation of a circle which has a radius of 3 and a centre located at $(2, -4)$.

Solution Substituting into $(x - a)^2 + (y - b)^2 = r^2$ gives

$$(x-2)^2+(y-(-4))^2 = 3^2$$
$$x^2 - 4x + 4 + y^2 + 8y + 16 = 9$$
$$x^2 - 4x + y^2 + 8y = -11$$

8.21 Calculate the radius and centre of the circle that is represented by the equation $x^2 - 4x + y^2 + 6y = 12$.

Solution The equation of the circle is

$$x^2 - 4x + y^2 + 6y = 12$$

Completing the squares gives

$$(x-2)^2 - 4 + (y+3)^2 - 9 = 12$$
$$(x-2)^2 + (y+3)^2 = 25$$

So the radius of the circle is 5 and the centre is at $(2, -3)$.

Transformation of a circle

In the previous section we saw that a circle is completely specified by the location of its centre and the length of its radius. Since the radius of a circle will be unaltered by either a translation or a rotation the effect of these transformations on a circle can be determined by considering how they modify the circle's centre.

Worked example

8.22 Determine the equation of the circle of radius 4 and centre (2,6) after it has been rotated through 90° about the point (4,8).

Solution The first task is to calculate the transformation matrix. The matrix which translates the origin to (4,8) is

$$\begin{bmatrix} 1 & 0 & 0 \\ 0 & 1 & 0 \\ -4 & -8 & 1 \end{bmatrix}$$

The rotation through 90° is given by

$$\begin{bmatrix} 0 & 1 & 0 \\ -1 & 0 & 0 \\ 0 & 0 & 1 \end{bmatrix}$$

The translation of the origin back to (0,0) is given by

$$\begin{bmatrix} 1 & 0 & 0 \\ 0 & 1 & 0 \\ 4 & 8 & 1 \end{bmatrix}$$

So the transformation matrix

$$\boldsymbol{T} = \begin{bmatrix} 1 & 0 & 0 \\ 0 & 1 & 0 \\ -4 & -8 & 1 \end{bmatrix} \begin{bmatrix} 0 & 1 & 0 \\ -1 & 0 & 0 \\ 0 & 0 & 1 \end{bmatrix} \begin{bmatrix} 1 & 0 & 0 \\ 0 & 1 & 0 \\ 4 & 8 & 1 \end{bmatrix}$$

$$= \begin{bmatrix} 0 & 1 & 0 \\ -1 & 0 & 0 \\ 12 & 4 & 1 \end{bmatrix}$$

Transforming the centre

$$[x^* \quad y^* \quad 1] = [2 \quad 6 \quad 1] \begin{bmatrix} 0 & 1 & 0 \\ -1 & 0 & 0 \\ 12 & 4 & 1 \end{bmatrix}$$

$$= [6 \quad 6 \quad 1]$$

$$(x^*, y^*) = (6, 6)$$

So the centre of the circle is now located at (6,6) and its radius is 4. The equation of this circle is:

$$(x-6)^2 + (y-6)^2 = 16$$
$$x^2 - 12x + y^2 - 12y = -56$$

Self-Assessment Questions 8.5

At what points does the circle with radius r and centred at the origin cut the x and y axes? What geometric shape is created by joining these points together?

Exercise 8.5

1. Write down the equation of the circle that has a radius of 4 and is centred on the origin.
2. Calculate the distance between the centre of the circle $x^2 - 6x + y^2 - 8y = 24$ and the origin.
3. Calculate the equation of the circle with a radius of 2 centred on (4,4) after it has been rotated through 45° about the origin.

Test and Assignment Exercises 8

1. Show that the point (4,4) is equidistant from the points (1,0) and $(-1, 4)$.
2. Write down the transformation matrix that represents a transformation consisting of a horizontal translation of 4 followed by a vertical translation of -7. Use this matrix to calculate the new co-ordinates of the point located at $(-4, 6)$ after it has been subjected to this transformation.
3. Calculate the new co-ordinates of the point $(3, -2)$ after it has been rotated through an angle of 45° about the origin.
4. What translation has the same effect as rotating the point (6,4) through 90° about the origin?
5. Calculate the transformation matrix that represents a rotation through 90° about the point (1,6). Use this matrix to enable you to sketch the straight line that passes through the points (2,4) and $(-3, 1)$ both before and after it has been subjected to the above rotation.
6. Derive the equation of the straight line that cuts the x axis at (b,0) and the y axis at (0,a).
7. Derive the equation of the straight line that is parallel to the line $2y - 4x + 6 = 0$ and passes through the point (5,3).
8. Calculate the gradient of the straight line that has the equation $3y + 6x - 12 = 0$. Calculate also the points at which this line cuts the x axis and y axis and produce a sketch of the line.
9. If two straight lines with the equations $y = m_1x + c$ and $y = m_2x + b$ intersect each other at right angles show that $m_1m_2 = -1$.
10. What happens to the co-ordinates of the point (x,y) that lie on a given straight line, when the line is rotated through 180° about a point (x_1, y_1) that also lies on the line?
11. Write down the equation of the circle that has a centre located at (2,3) and a radius of 4. What are the co-ordinates of the points at which this circle cuts the x axis and the y axis?
12. Prove that the location of any circle remains unchanged when it is rotated through any angle about its centre.

8.6 Further reading

Donald Hearn and M. Pauline Baker, *Computer Graphics* (Prentice-Hall International, 1994)

This text provides an introduction to the representation and manipulation of graphical images in two and three dimensions.

Foley, van Dam, Feine and Hughes, *Computer Graphics Principles and Practice* (Addison-Wesley, 1990)

A comprehensive text which not only covers the fundamental mathematical elements but deals with such topics as solid modelling and colouring techniques.

S G Hoggar, *Mathematics for Computer Graphics* (Cambridge University Press, 1992)

A more advanced text than the previous two which explores a variety of mathematical techniques for representing computer graphics. As well as dealing with matrices and vectors it also looks at fractals and topology.

9 Calculus

Objectives

When you have mastered this chapter you should be able to

- understand what is meant by, and be able to calculate, average rates of change
- understand the relationship between average rate of change, instantaneous rate of change and derivative
- calculate the derivative of a simple function from first principles
- apply the rules of differentiation to simple combinations of functions
- use differentiation to identify and classify the stationary points of a function
- use the identification of stationary points to solve simple optimization problems
- find definite and indefinite integrals of simple functions and use them to find areas

This chapter provides a very basic introduction to the **differential calculus** and the **integral calculus**. The differential calculus is concerned with how quantities change, and provides mathematical techniques for modelling and analysing dynamic systems. It can be employed in such diverse areas as computer animation, the modelling of performance and costs of computer projects, the analysis of software reliability and the simulation of computer systems. The integral calculus is concerned with how things accumulate, and is used in various situations to do with modelling, when we need for example to find averages, in image processing, where Fourier transforms give a way of filtering and manipulating signals, and in many other application areas. We will not examine these applications in detail in this chapter, but instead will concentrate on obtaining an understanding of the basic ideas that can be applied to any of the appropriate areas.

9.1 Average rate of change

Imagine a point moving along a straight line. We will denote its position – *i.e.* the distance it is to the right of some reference point – at time t by $p(t)$. See Figure 9.1. If position is related to time by $p(t) = t^2$, then the following are some corresponding values for these two variables:

Time in seconds (t)	1	2	3	4	5	6
Position in metres $(p(t))$	1	4	9	16	25	36

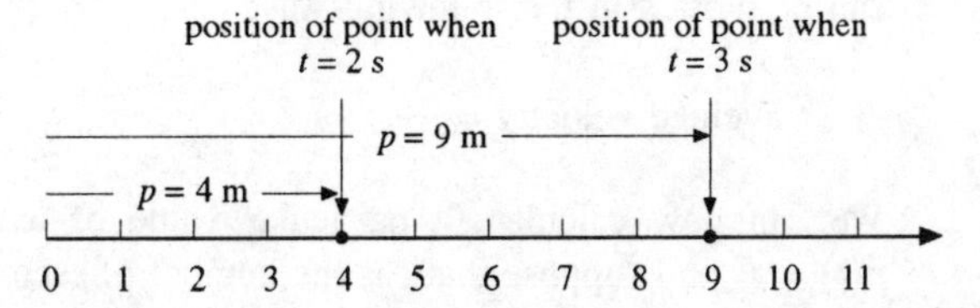

Figure 9.1
A point moving along a straight line

We can use these corresponding pairs of values as co-ordinate pairs to create a graph representing the point as shown in Figure 9.2. As our point moves we expect it to have a velocity at any given time, just as we do when travelling along the road in a car; at any given time, we can look at the speedometer and see what our current velocity is. But what does this mean?

It is easy to work out the average velocity over a given time interval. If the point is at p_1 at time t_1 and at p_2 at time t_2, then the average velocity of the point over that interval, which is by definition the displacement divided by the

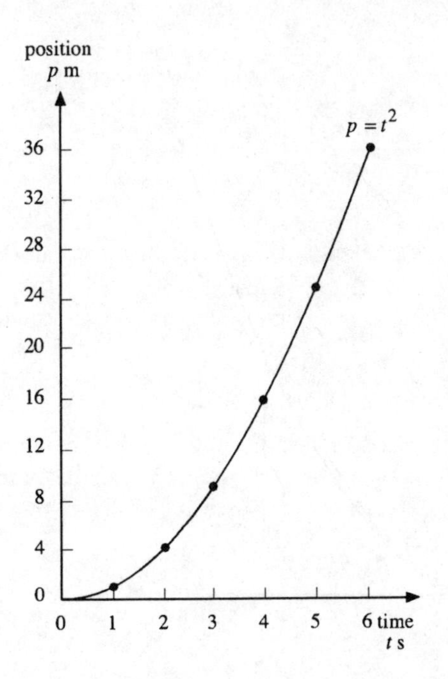

Figure 9.2
Graph of position against time for a moving point

time taken, is given by

$$\text{average velocity} = \frac{\text{distance travelled}}{\text{time taken}} = \frac{p_2 - p_1}{t_2 - t_1}$$

Since p is given by $p(t)$, we could also write this as

$$\frac{p(t_2) - p(t_1)}{t_2 - t_1}$$

But there is also a standard notation used to make this rather easier to write. We use Δt to denote $t_2 - t_1$; the Greek letter Δ is used to denote a change. The corresponding change in p is denoted Δp, so the average velocity of the point can be written in the following way:

$$\text{average velocity} = \frac{\Delta p}{\Delta t}$$

We can now calculate a particular value of average velocity; recall that $p(t) = t^2$ and suppose that t is the number of seconds that have passed, while $p(t)$ is the number of metres the point is to the right of the reference point after t seconds. Then when $t = 2$, $p(t) = 4$, and when $t = 4$, $p(t) = 16$. These values give

$$\frac{\Delta p}{\Delta t} = \frac{16 - 4}{4 - 2} = \frac{12}{2}$$

so that the average velocity over this time interval is 6 metres per second.

From Figure 9.3 we see that the average velocity of the point between the times $t = 2$ and $t = 4$ is just the gradient of the line joining the points $(2, 4)$ and $(4, 16)$. In fact it is easy to work out the average velocity for any time

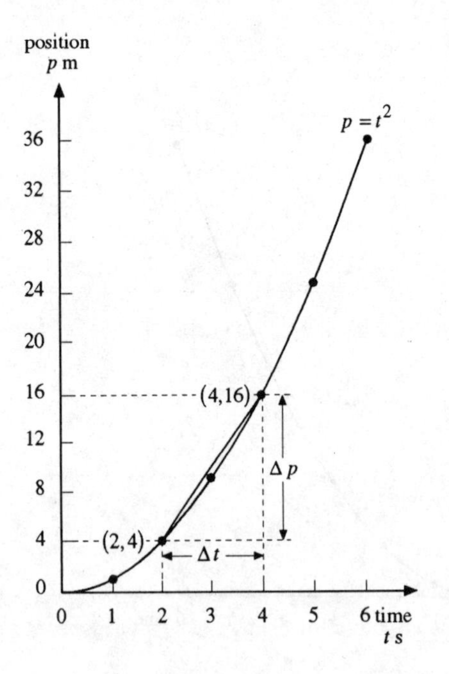

Figure 9.3
The gradient of the line joining (2, 4) and (4, 16) is equal to $\Delta p/\Delta t$

interval. Let the interval begin after t_1 seconds and end after t_2 seconds. Then $p(t_1) = t_1^2$ and $p(t_2) = t_2^2$, so that

$$\frac{\Delta p}{\Delta t} = \frac{t_2^2 - t_1^2}{t_2 - t_1} = \frac{(t_2 + t_1)(t_2 - t_1)}{t_2 - t_1} = t_2 + t_1$$

so for any values of t_1 and t_2, the average velocity of the point between t_1 and t_2 seconds is $t_1 + t_2$ metres per second. Again, we see, now in Figure 9.4, that this is the gradient of a line, this time the line joining the points (t_1, t_1^2) and (t_2, t_2^2).

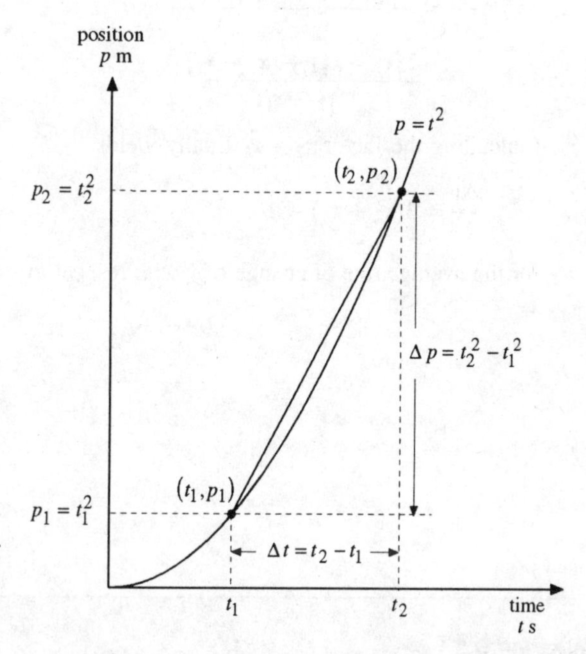

Figure 9.4
The average velocity of a point between times t_1 and t_2 is equal to the gradient of the line joining (t_1, p_1) and (t_2, p_2)

When p represents position and t represents time, we think of the quantity we have been calculating as the average velocity; but if we are given any two quantities, say x and y, where the value of y depends on (and only on) the value of x, we can do much the same thing. We can find $\Delta y/\Delta x$, and this quantity is called the **average rate of change of y with respect to x.**

Worked example

9.1 If two variables x and y are related by the equation $y = 3x^2 - x$ calculate

1. the average rate of change of y with respect to x as x varies from 2 to 4
2. the average rate of change of y with respect to x as x varies from x_1 to x_2

Solution 1. The average rate of change of y with respect to x is given by

$$\frac{\text{change of } y}{\text{change of } x} = \frac{\Delta y}{\Delta x}$$

Since $y = 3x^2 - x$, we see that when $x = 4$, $y = 3 \times 4^2 - 4 = 44$ and when $x = 2$, $y = 3 \times 2^2 - 2 = 10$, and so the average rate of change is

$$\frac{\Delta y}{\Delta x} = \frac{44 - 10}{4 - 2} = 17$$

2. For the more general case we still have $y = 3x^2 - x$, and so we obtain

$$\begin{aligned}\frac{\Delta y}{\Delta x} &= \frac{(3x_2^2 - x_2) - (3x_1^2 - x_1)}{x_2 - x_1} \\ &= \frac{3(x_2^2 - x_1^2) + (x_1 - x_2)}{x_2 - x_1}\end{aligned}$$

Cancelling the factor $x_2 - x_1$ finally yields

$$\frac{\Delta y}{\Delta x} = 3(x_2 + x_1) - 1$$

for the average rate of change of y with respect to x as x varies from x_1 to x_2.

Self-Assessment Question 9.1

If y is a function of x, what geometric quantity is the average rate of change of y with respect to x as x increases from 0 to 1?

Exercise 9.1

1. If the variables x and y are related by $y = 4 - 6x$ find the average rate of change of y with respect to x as x varies from 3 to 3.5.
2. If the variables s and t are related by $s = \sqrt{t + 12}$, find the average rate of change of s with respect to t as t varies from 24 to 52.
3. If w and v are related by $w = v^2 + 2v$, find the average rate of change of w with respect to v as v varies from c to $c + h$.
4. The number of staff, n, working on a computer project is related to the time that the project has been running, t, measured in weeks, by the relationship $n = 40t - 6t^2$. Find the average rate of change in staffing for the project between weeks 2 and 4.

9.2 Instantaneous rate of change

In the previous section we saw how to calculate the average rate of change of one quantity with respect to another. But there are times when we wish to know the rate of change of one quantity with respect to another at some particular value of the latter; as when we want to know our velocity at a given time while driving, so as to be sure that we are not exceeding the speed limit. The problem is that velocity is defined to be the ratio of displacement to time taken; at a given instant, no time has passed, so what does velocity at that instant mean?

The way we try to approach this is to find the average velocity over a very small time interval near the time we are interested in. If, as we make the interval smaller and smaller, the average velocity approaches some particular value, we regard that value as the instantaneous velocity.

Let us consider what happens in the case where $p(t)$ is the position of a moving point at time t. We will consider a small time interval starting at time t. By convention, the small time interval runs from time t to time $t + \delta t$. Just as Δ was used to represent a change, δ is used to represent a small change. So the length of the small time interval is δt. The change in p over this time interval is $p(t + \delta t) - p(t)$ which is denoted by δp; so δp is the change in p that occurs over the time interval of length δt. So the average velocity over that time interval is just

$$\frac{p(t+\delta t) - p(t)}{(t+\delta t) - t} = \frac{\delta p}{\delta t}$$

What we have to do is find out what happens when δt is very small indeed.

So, let us see what happens when $p(t) = t^2$, the special example we considered first in Section 9.1. In this case we have

$$\begin{aligned}\frac{\delta p}{\delta t} &= \frac{(t+\delta t)^2 - t^2}{(t+\delta t) - t} \\ &= \frac{t^2 + 2t\delta t + (\delta t)^2 - t^2}{\delta t} \\ &= \frac{2t\delta t + (\delta t)^2}{\delta t} \\ &= 2t + \delta t\end{aligned}$$

as long as δt is not actually equal to zero. This is illustrated in Figure 9.5, where we see that the average velocity is the gradient of the line joining the points $(t + \delta t, (t + \delta t)^2)$ and (t, t^2). If we now make δt very small, the value of the average velocity gets closer and closer to $2t$; in fact we can make it as close as we want by making δt sufficiently small. The mathematical expression

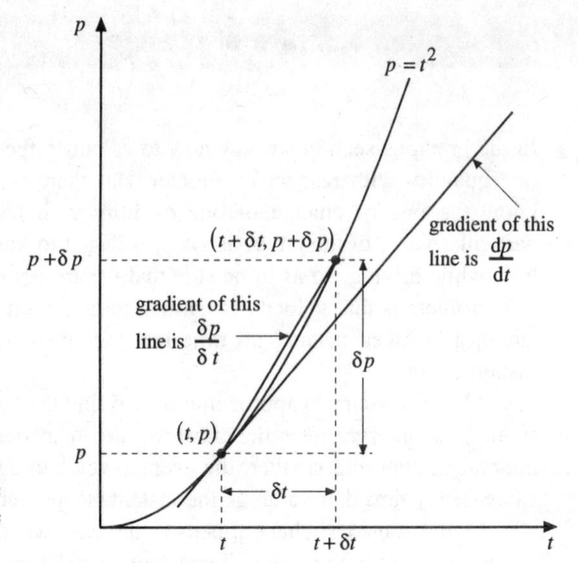

Figure 9.5
The gradient of a tangent to a curve is equal to the value of the derivative of the curve evaluated at that point

used to express this state of affairs is that $2t$ is the **limit** as δt tends to 0 of $\delta p/\delta t$. We write

$$\lim_{\delta t \to 0} \frac{\delta p}{\delta t} = \lim_{\delta t \to 0} 2t + \delta t = 2t$$

This new idea is described by new notation; we write

$$\lim_{\delta t \to 0} \frac{\delta p}{\delta t} = \frac{dp}{dt}$$

and call this quantity the **derivative** of p with respect to t. It is what we mean by the instantaneous rate of change of p with respect to t, or the instantaneous velocity of the point at time t. From now on, 'rate of change' will mean 'instantaneous rate of change' unless we say otherwise. Thus in this case, we have the relationships between time, position and instantaneous velocity given by

$$p(t) = t^2$$
$$\frac{dp}{dt} = 2t$$

Note that the existence of the derivative of p is not guaranteed; the limit will exist for some functions, and not for others. However, the derivative will exist for all the functions you meet in this text. Discussion of this point can be found in the suggested material for further reading at the end of the chapter.

If you look again at Figure 9.5 you will see two straight lines shown on it; one is the segment connecting points with coordinates $(t + \delta t, (t + \delta t)^2)$ and (t, t^2), and its gradient is $2t + \delta t$. The other is the line passing through the point (t, t^2) with gradient $2t$. This second line is called the **tangent** to the graph at the point (t, t^2). It is the straight line that stays closest to the curve in a small neighbourhood of the point (t, t^2).

The process we have outlined above is called **differentiation**; the calculation that found the derivative of p in the special case where $p(t) = t^2$ is called **differentiation from first principles**.

It is useful at this point to recapitulate what we have done here, for the case $p(t) = t^2$.

$$\begin{aligned}
\frac{dp}{dt} &= \lim_{\delta t \to 0} \frac{p(t+\delta t) - p(t)}{(t+\delta t) - t} \\
&= \lim_{\delta t \to 0} \frac{(t+\delta t)^2 - t^2}{\delta t} \\
&= \lim_{\delta t \to 0} \frac{t^2 + 2t\delta t + (\delta t)^2 - t^2}{\delta t} \\
&= \lim_{\delta t \to 0} \frac{2t\delta t + (\delta t)^2}{\delta t} \\
&= \lim_{\delta t \to 0} 2t + \delta t \\
&= 2t
\end{aligned}$$

We can differentiate other functions by following the same scheme, as we will see. Note that it does not matter what letters are used to represent the quantities involved, the process is the same on each occasion. If we have two variables, say x and y, and the value of y is determined by some formula involving x, we often call y the **dependent** variable (since its value depends on that of x) and x the **independent** variable.

Worked examples

9.2 If y and x are related by $y = 2x^3 - 3x$, then use differentiation from first principles to find the derivative of y with respect to x.

Solution Since $y = 2x^3 - 3x$ we have

$$\begin{aligned}
\frac{\delta y}{\delta x} &= \frac{y(x+\delta x) - y(x)}{\delta x} \\
&= \frac{2(x+\delta x)^3 - 3(x+\delta x) - (2x^3 - 3x)}{(x+\delta x) - x} \\
&= \frac{2(x^3 + 3x^2\delta x + 3x(\delta x)^2 + (\delta x)^3) - 3(x+\delta x) - (2x^3 - 3x)}{\delta x} \\
&= \frac{2x^3 + 6x^2\delta x + 6x(\delta x)^2 + 2(\delta x)^3 - 3x - 3\delta x - 2x^3 + 3x}{\delta x} \\
&= \frac{6x^2\delta x + 6x(\delta x)^2 + 2(\delta x)^3 - 3\delta x}{\delta x} \\
&= 6x^2 - 3 + 6x(\delta x) + 2(\delta x)^2
\end{aligned}$$

This tells us the average rate of change of y with respect to x over the range x to $x + \delta x$; to find out the rate of change at x we have to see what happens when we make δx very small. First, we consider the term $6x\delta x$. It does not matter how large $6x$ is, by making δx sufficiently small we can make $6x\delta x$ as small as we want. Also, $(\delta x)^2$ is even smaller than δx. So both the terms with δx in them are negligibly small if δx is small enough. We thus have

$$\frac{dy}{dx} = \lim_{\delta x \to 0} \frac{\delta y}{\delta x} = 6x^2 - 3$$

9.3 The volume in cubic metres, V, of a sphere is related to its radius in metres, r, by $V = \frac{4}{3}\pi r^3$. If a spherical balloon is being inflated, find the rate at which the volume is changing with respect to the radius when $r = 2$.

Solution We have $V = \frac{4}{3}\pi r^3$, so that

$$\begin{aligned}\frac{\delta V}{\delta r} &= \frac{\frac{4}{3}\pi(r+\delta r)^3 - \frac{4}{3}\pi r^3}{(r+\delta r) - r} \\ &= \frac{4}{3}\pi\frac{(r+\delta r)^3 - r^3}{\delta r} \\ &= \frac{4}{3}\pi\frac{r^3 + 3r^2\delta r + 3r(\delta r)^2 + (\delta r)^3 - r^3}{\delta r} \\ &= \frac{4}{3}\pi\frac{3r^2\delta r + 3r(\delta r)^2 + (\delta r)^3}{\delta r} \\ &= \frac{4}{3}\pi(3r^2 + 3r\delta r + (\delta r)^2)\end{aligned}$$

As before, when we make δr very small, each of the terms involving δr vanishes and we are just left with

$$\frac{dV}{dr} = 4\pi r^2$$

Thus, when $r = 2$, the rate of change of V with respect to r is 16π.

Self-Assessment Questions 9.2

What is the relation between dV/dr and the surface area for a sphere? Do you think it is a coincidence?

Exercise 9.2

1. Find dy/dx for each of the following cases:

   (i) $y = 2x$ (ii) $y = 4x^2$ (iii) $y = 2x^3$

2. The area, A, of a circle of radius r is πr^2. If the radius of a circle being displayed on a computer screen is gradually increasing use differentiation from first principles to find the rate of change of area with respect to radius when the radius is 4 units in length.

9.3 What does the derivative tell us?

We can use the derivative in two different ways, to tell us about the way that two related quantities vary together; one is qualitative, the other quantitative.

First, we consider the quantitative information that we can obtain from a derivative. Suppose that we have two quantities, x and y, and that y is determined by some formula involving x. If we pick a very small change in x, say δx, then we know that

$$\frac{\delta y}{\delta x} \approx \frac{dy}{dx}$$

and so we can deduce that

$$\delta y \approx \frac{dy}{dx} \delta x$$

In other words, we find that the change in y is approximately given by the change in x multiplied by the derivative of y with respect to x.

Worked example

9.4 Compare the exact change in the volume of a sphere to the change given by this approximation if its radius increases from 4 metres to 4.1 metres; also when the radius decreases from 4 metres to 3.99 metres.

Solution From Worked example 9.3 we know that $V = \frac{4}{3}\pi r^3$, so when $r = 4$, $V = \frac{256}{3}\pi \approx 268.08$ cubic metres. When $r = 4.1$, we get $V \approx 288.70$ cubic metres, so the change in volume is 20.62 cubic metres. We also know that $dV/dr = 4\pi r^2$, which is 64π when $r = 4$, and multiplying this by the change in r gives $64\pi \times 0.1 \approx 20.11$ cubic metres. This is fairly close, with an error of about 2.5%.

When $r = 3.99$, we get $V \approx 266.08$ cubic metres, so the change in volume is approximately -2 cubic metres, *i.e.* the volume decreases by approximately 2 cubic metres. The approximation obtained using the derivative is now $dV/dr \times -0.01 = 64\pi \times -0.01 \approx -2.01$ cubic metres.

We see that the approximation is much better for the smaller change; this is generally the case. You get a better approximation to the change in the dependent quantity by considering a smaller change in the quantity it depends on.

Now that we have seen how to use the derivative of a function to find quantitative information, we can see how it also gives qualitative information. If the derivative is positive, then the ratio of the change in the independent variable to the change in the dependent variable is positive, *i.e.* they must both be the same sign. This means that when we increase the independent variable by a small amount, the dependent one also increases. If the derivative is negative, then increasing the independent variable causes the dependent one to decrease.

Worked example

9.5 Does $y = 2x^3 - 3x$ increase or decrease when x is increased slightly from 2? What if x is decreased slightly from 0?

Solution From Worked example 9.2 we know that in this case $dy/dx = 6x^2 - 3$, so when $x = 2$ the derivative is $24 - 3 = 21$. Thus if x is increased the change in x is positive, and multiplying by a positive derivative we get a positive change – *i.e.* an increase – in y. On the other hand, when $x = 0$, the derivative is -3. But if x is decreased, the change in x is negative and multiplying this by a negative derivative again gives a positive change, *i.e.* an increase in y.

Self-Assessment Question 9.3

What do you think happens if the derivative of a function is 0 at some value of the independent variable? You may find it useful to consider the functions given by $y = x^2$ and $y = x^3$ and the value $x = 0$.

Exercise 9.3

1. If $y = 4x^2$, find the exact change and the approximate change when x is increased from 2 to 2.1, 2.01 and 2.001. Find the percentage error in the approximate change in each case. When is the percentage error smallest? (Note: you found the derivative of this function in Exercise 9.2.)
2. If $y = 2x^3 - 3x$, work out whether y increases if x is (i) increased slightly from 0 and (ii) decreased slightly from 4.

9.4 Differentiating simple functions

So far, whenever we wanted to know a derivative, we had to work it out from first principles. This is a time consuming business, and you might quickly come to suspect that you must be re-inventing the wheel to a considerable extent every time you do this. You would be right: the derivatives of certain standard functions are known, as are some rules for differentiating most kinds of combinations of functions. We will only consider functions which are of one simple type; but the basic principles are the same for the other functions that one can differentiate.

All the functions we will consider are those that can be constructed from powers of the independent variable. We have our first rule:

KEY POINT

Rule 1 If $y = kx^n$, where k is any constant, then $dy/dx = nkx^{n-1}$.

This rule works for any value of n, whether it is positive, negative, a whole number, a fraction, or any other number. We can use this rule on any function of the appropriate type, without having to go through the tedious business of finding limits explicitly. The rule also works on a special case that could easily be missed; if $n = 0$, we have $y = kx^0 = k$, and in this case the derivative is 0.

Worked example

9.6 Using Rule 1, differentiate each of

(i) $y = 4x^8$ (ii) $y = \dfrac{3}{x^2}$ (iii) $y = 2.4\sqrt{x}$ (iv) $y = \dfrac{6}{\sqrt{x^3}}$

with respect to x.

Solution In each case, we write the function in the form $y = kx^n$ and apply the rule.

(i) $y = 4x^8$ is in the required form. The derivative is therefore

$$\frac{dy}{dx} = 4 \times 8x^{8-1} = 32x^7$$

(ii) $y = \dfrac{3}{x^2} = 3x^{-2}$ and so

$$\frac{dy}{dx} = 3 \times (-2)x^{-2-1}$$

$$= -6x^{-3} = \frac{-6}{x^3}$$

(iii) $y = 2.4\sqrt{x} = 2.4x^{0.5}$ and so

$$\frac{dy}{dx} = 2.4 \times 0.5x^{0.5-1}$$

$$= 1.2x^{-0.5} = \frac{1.2}{\sqrt{x}}$$

(iv) $y = \dfrac{6}{\sqrt{x^3}} = 6x^{-1.5}$, and so

$$\frac{dy}{dx} = 6 \times (-1.5)x^{-1.5-1}$$

$$= -9x^{-2.5} = \frac{-9}{\sqrt{x^5}}$$

Thus Rule 1 enables us to differentiate any function that has the form *dependent variable is a multiple of a power of the independent variable.* If we knew how to differentiate any sum of such functions, we would be in a position to deal with a collection of functions large enough to be useful for the modelling and analysis of interesting situations.

KEY POINT

Rule 2 The derivative of the sum of two functions is the sum of the derivatives of the two functions.

And, since any sum can be made up by adding things in pairs, this rule is enough to enable us to find the derivative of any sum of multiples of powers.

Worked example

9.7 Differentiate each of the following with respect to the independent variable.
(i) $y = 2x^3 - 3x^2 + 5x$ (ii) $p = 2t - \frac{1}{t^2}$ (iii) $w = 2\sqrt{z} - \frac{2}{z}$ (iv) $Q = \frac{2w^2 - w}{w^3}$

Solution Again, we convert each into a sum of multiples of powers and use the rules we have.

(i) This one is already in the required form. We can write down

$$\frac{dy}{dx} = 6x^2 - 6x + 5$$

(ii) First, we rearrange the formula to get it into the form we require,

$$p = 2t - \frac{1}{t^2} = 2t - t^{-2}$$

From this it follows that

$$\frac{dp}{dt} = 2 - (-2)t^{-3} = 2 + 2t^{-3} = 2 + \frac{2}{t^3}$$

(iii) Again, we have to re-write the expression for w,

$$w = 2\sqrt{z} - \frac{2}{z} = 2z^{0.5} - 2z^{-1}$$

Thus

$$\frac{dw}{dz} = z^{-0.5} + 2z^{-2} = \frac{1}{\sqrt{z}} + \frac{2}{z^2}$$

(iv) This one looks as if it cannot be done with the rules we have, but a little algebra converts it to the familiar form.

$$Q = \frac{2w^2 - w}{w^3} = \frac{2}{w} - \frac{1}{w^2} = 2w^{-1} - w^{-2}$$

and so

$$\frac{dQ}{dw} = -2w^{-2} + 2w^{-3} = \frac{2}{w^3} - \frac{2}{w^2}$$

So we see that the actual calculation of derivatives is a very straightforward, even mechanical, task. Once the basic rules are grasped, the process of finding the derivative of any function in the family we have considered is simply a matter of using the basic laws of indices.

We can also use these rules to enable us to find various other quantities of interest.

Worked examples

9.8 A curve is described by the equation $y = 2x^3 - 3x$. Find the gradient of the tangent to this curve at $(2, 10)$.

Solution Since $y = 2x^3 - 3x$, we can immediately see that $dy/dx = 6x^2 - 3$. When $x = 2$, this gives 21, and so the gradient of the tangent at that value of x is 21.

9.9 A point is moving in such a way that p, the distance in metres it lies to the right of some reference point, is given by $\frac{1}{2}t^2 + 2t$, where t is the number of seconds that have elapsed. Find its velocity when $t = 3$.

Solution Since $p = \frac{1}{2}t^2 + 2t$, we immediately note that $dp/dt = t + 2$. Thus when $t = 3$, the velocity is 5 metres per second.

Self-Assessment Question 9.4

You now know that the derivative of the sum of two functions is the sum of their derivatives. Find an example to show that the derivative of the product of two functions is not the product of their derivatives.

Exercise 9.4

1. In each of the following cases, work out the derivative you are asked for:

   (i) $y = 6x^{-1}$, find $\dfrac{dy}{dx}$ (ii) $A = 6r^2$, find $\dfrac{dA}{dr}$

   (iii) $x = \dfrac{3}{\sqrt{y}}$, find $\dfrac{dx}{dy}$

2. (a) Differentiate each of the following with respect to the independent variable.

   (i) $y = 2x^2 - 3x + 6$ (ii) $s = 2t + 5t^2$

   (iii) $y = \dfrac{2}{x^2} - \sqrt{x}$

   (iv) $v = 4t + \dfrac{1}{2\sqrt{t}}$

   (b) Find the value of the gradient to the curve $y = x^2 - x^3$ at the point where the x coordinate has a value of 2.

   (c) Show that the tangent to the curve $y = x^2 + 2x - 3$ at the point $(-2, -3)$ is parallel to the line given by $2y + 4x - 3 = 0$.

   (d) A software company has just released a new product onto the market. Their analysts predict that the monthly sales Q will be given by $400m - 10m^2$, where m is the number of months since the launch, for the first 40 months. Use differentiation to show that sales are increasing after 10 months, but decreasing after 30 months.

9.5 Stationary points and optimization

Let us suppose that some quantity y in which we are interested depends on some other quantity x, and we want to know at what value of x in some interval y takes its largest value. There are three possibilities as to what can happen: y may be largest just at the start of the interval, just at the end or at some point in between. We will consider what must happen if the largest value happens at neither end of the interval.

Now, y is given by some expression in x, so we can think about the derivative $\mathrm{d}y/\mathrm{d}x$. First, note that $\mathrm{d}y/\mathrm{d}x$ cannot be positive when x is at its largest value, for if the derivative is positive we can make y larger by increasing x a little. Similarly, the derivative cannot be negative, for then we could make y larger by decreasing x by a little. The only remaining possibility is that the derivative is 0 when y takes on its largest value. Furthermore, the function must have been increasing just before this, so the derivative should be positive for slightly smaller x; also y should be decreasing just after, and so the derivative should be negative for slightly larger x.

This picture lets us pick values which are candidates for when y is at its largest value. We simply consider all those values of x that make $\mathrm{d}y/\mathrm{d}x$ change sign from positive to negative, and then see at which of them y is largest; we should also check the very beginning of the interval we are interested in and the very end, just in case y is actually largest at either of those positions. Then we know when the largest value of y occurs.

By just the same form of argument, y must take its smallest value either at the beginning or the end of the interval of interest, or where $\mathrm{d}y/\mathrm{d}x$ is zero and is changing sign from negative to positive.

Worked example

9.10 Find the values of x that make y largest and smallest when x lies between -1 and 1.4, and y is given by $y = 2x^3 - 3x^2 + 2$. This curve is sketched in Figure 9.6.

Solution First, we note that $\mathrm{d}y/\mathrm{d}x = 6x^2 - 6x = 6x(x-1)$ and so is zero when $x = 0$ or $x = 1$. If x is approximately 0, the derivative is given approximately by $-6x$, and so when x increases from a value slightly below 0 to one slightly above 0, the derivative changes from positive to negative. Therefore this could be when y takes its greatest value. If x is approximately 1, then the derivative is approximately $6(x-1)$, and so when x increases from just below 1 to just above 1, the derivative changes sign from negative to positive. Hence this could be where y takes its smallest value.

Now, when $x = -1$, $y = -3$ and when $x = 1.4$, $y = 1.608$. When $x = 0$, $y = 2$, so this is the largest value y takes. When $x = 1$, $y = 1$, so the smallest value y takes is -3, when $x = -1$.

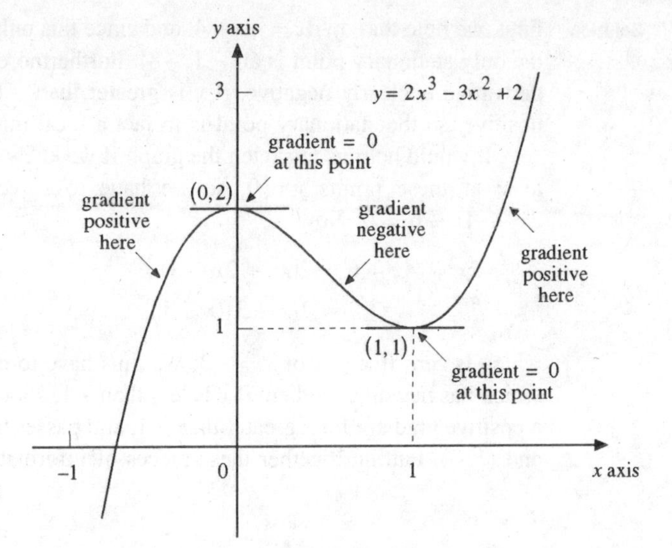

Figure 9.6
Sketch of the curve $y = 2x^3 - 3x^2 + 2$ showing change in gradient

The points on the graph of a function where the derivative of the function vanishes are its **stationary points**. If the derivative actually changes sign, we call the points **turning points**, and more precisely, a point is a **local minimum** if the derivative changes from negative to positive, and a **local maximum** if it changes from positive to negative. If the sign of the derivative is the same at both sides of the stationary point, we call it a **saddle point**. Figure 9.7 shows a selection of stationary points on a graph.

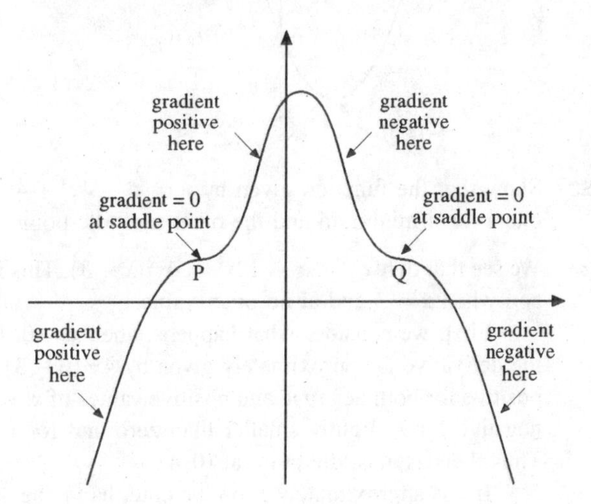

Figure 9.7
A sketch graph showing saddle points

Worked examples

9.11 Use differentiation to find the stationary point on the graph of $y = 2x^2 + 4x - 6$. Show that this stationary point is in fact a local maximum and hence draw a rough sketch of the graph.

Solution First, we note that $dy/dx = 4x + 4$, and since this only vanishes when $x = -1$, the only stationary point is at $(-1, -8)$. Furthermore, if x is less than -1, the derivative is clearly negative, if x is greater than -1 the derivative is clearly positive, so the stationary point is in fact a local minimum.

It would help us to sketch the graph if we knew where the graph cut the x axis; at these points $y = 0$, so we have to solve the quadratic equation $2x^2 + 4x - 6 = 0$. Now,

$$\begin{aligned} 2x^2 + 4x - 6 &= 2(x^2 + 2x - 3) \\ &= 2(x + 3)(x - 1) \end{aligned}$$

which is zero if $x = 1$ or $x = -3$. We thus have to draw the graph of a curve which has negative gradient if x is less than -1, a local minimum at $(-1, -8)$, a positive gradient for x greater than -1, and passes through the points $(-3, 0)$ and $(1, 0)$. Putting together these pieces of information gives us Figure 9.8.

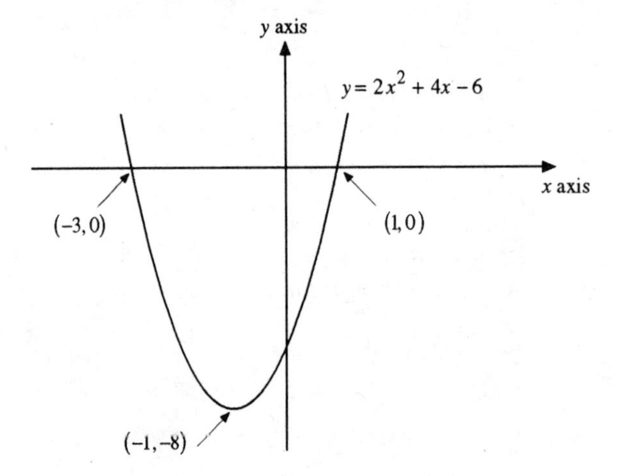

Figure 9.8
A sketch graph of the function $y = 2x^2 + 4x - 6$

9.12 Show that the function given by $y = x^4 - 4x^3 + 4$ has two stationary points, one a local minimum and the other a saddle point.

Solution We see that $dy/dx = 4x^3 - 12x^2 = 4x^2(x - 3)$. This is clearly zero when $x = 0$ and when $x = 3$ and at no other value.

First, we consider what happens when $x = 0$. For values of x near zero, the derivative is approximately given by $4x^2(0 - 3) = -12x^2$, and since x^2 is positive for both negative and positive values of x, we see that the derivative is negative for x slightly smaller than zero and for x slightly larger than zero. Thus there is a saddle point at $(0, 4)$.

If x is approximately 3, on the other hand, the derivative is approximately $4(3^2)(x - 3) = 36(x - 3)$, which is negative if x is slightly less than 3 but positive if x is slightly greater than 3. Thus there is a local minimum at $(3, -23)$.

Note that in these cases it has been possible to work out the behaviour of the derivative near stationary points by inspecting the form of the derivative.

Sometimes this is more awkward, and you may find it helpful simply to evaluate the derivative for values of the independent variable near the stationary value.

To summarize, if we are given the relationship $y = f(x)$ between x and y, where $f(x)$ is some expression involving x, a point $(x, f(x))$ is a stationary point if the derivative vanishes there. Furthermore, it is a local maximum if the derivative changes sign from positive to negative as x increases, a local minimum if the derivative changes sign from negative to positive, and a saddle point otherwise.

If we wish to find the largest value y takes for some range of values of x, this can only happen at a local maximum or at an end of the range; if we wish the smallest value, this can only happen at a local minimum or at an end of the range.

Note: It is possible to approach the problem of finding out what kind a stationary point is by using other methods, in particular by differentiating again. This technique can be found in books that consider the calculus in more detail than we will here.

We can now look at an example involving some modelling.

Worked example

9.13 As a marketing ploy a manufacturer has decided to house a new computer in a cylindrical case. If the case must have a capacity of 16 litres, determine the dimensions of the case that minimize its surface area, thus minimizing the amount of material required for its manufacture.

Solution Let r be the radius of the cylindrical case and l be its length. Then the volume V is given by $V = \pi r^2 l$, and the surface area A is the area of the tube plus twice the area of an end, so $A = 2\pi r^2 + 2\pi r l$. Note that r must be a positive number.

Let us measure length in centimetres. Then the volume V is 16000 cubic centimetres, and so we have

$$16000 = \pi r^2 l \text{ which gives } l = \frac{16000}{\pi r^2}$$

Substituting this into the formula for surface area we find

$$\begin{aligned} A &= 2\pi r \frac{16000}{\pi r^2} + 2\pi r^2 \\ &= \frac{32000}{r} + 2\pi r^2 \end{aligned}$$

Next, we have to find any local minima of this function of r. Now,

$$\begin{aligned} \frac{dA}{dr} &= \frac{-32000}{r^2} + 4\pi r \\ &= \frac{4}{r^2}(\pi r^3 - 8000) \end{aligned}$$

and so the stationary points satisfy $\pi r^3 = 8000$ which has just one solution, when

$$r^3 = \frac{8000}{\pi}$$

and so

$$r = \frac{20}{\pi^{1/3}}$$

We need to know what kind of stationary point this gives; from the formula for the derivative we see that if r is less than $20\pi^{-1/3}$ the derivative is negative, while if r is greater than this the derivative is positive.

It follows that A increases from the values when $r = 20/\pi^{1/3}$ whether we increase or decrease r, and so this must be a local minimum. It is the only turning point, and A becomes extremely large if r is near zero or large, thus it must give the overall minimum.

We can now calculate l using the formula $l = 16000/\pi r^2$, giving $r \approx 13.7$ cm and $l \approx 27.3$ cm.

Thus to minimize surface area, a cylinder of volume 16 litres should have a radius of 13.7 cm and a height of 27.3 cm.

Self-Assessment Questions 9.5

Given a relationship $y = f(x)$, you can find the derivative of y with respect to x, and the sign of this quantity tells you whether y is increasing or decreasing. The derivative itself depends on x, and can be differentiated in turn. What does the sign of this second derivative tell you? What does the sign tell you if the first derivative happens to be zero?

Exercise 9.5

1. For each of the following, find the stationary points and classify them as local maximum, local minimum, or saddle point.
(a) $y = x^3 - x^2$ (b) $y = 2 - x^3$

2. The position p of a moving point at time t is given by $p = 64t - 16t^2$. By finding and analysing the stationary point of the graph of this function draw a sketch of this function, and show that the point is a maximum distance to the right of its starting point when $t = 2$.

3. Find the stationary points of the following functions and classify them as local maxima, local minima or saddle points.

(a) $y = x^4 - 4x^3 - 6$ (b) $y = x^3 - 3x^2 + 2$

(c) $p = 3v(1 - v^2)$ (d) $s = t + \frac{1}{t}$

4. The quantity of heat, H, generated by a piece of computing equipment is related to the amount of current, i, that it is using by the equation $H = 8i - i^2$. Determine the value of i that results in a maximum value of H for i between 0 and 6 and calculate this value of H.

5. The cost per user, C, in pounds sterling of a particular configuration is related to the number of simultaneous users, n, that the system can support by the equation

$$C = 6000 + \frac{180}{n} + \frac{n}{20}$$

By identifying the minimum turning point of this function calculate how many simultaneous users can be supported by the system that has the minimum cost per user.

9.6 Integration

So far we have been concerned with the differential calculus, which is relevant to questions about how fast quantities change with respect to one another. We will now take a very brief look at the integral calculus, which is concerned with questions about how quantities accumulate.

Again, let us consider a moving point whose position at time t is given by $p(t)$. Now, rather than trying to work out the speed of the point at a given time, let us consider the opposite problem. Suppose we know the point's speed $v(t)$ as a function of time, and want to find out how far it has travelled.

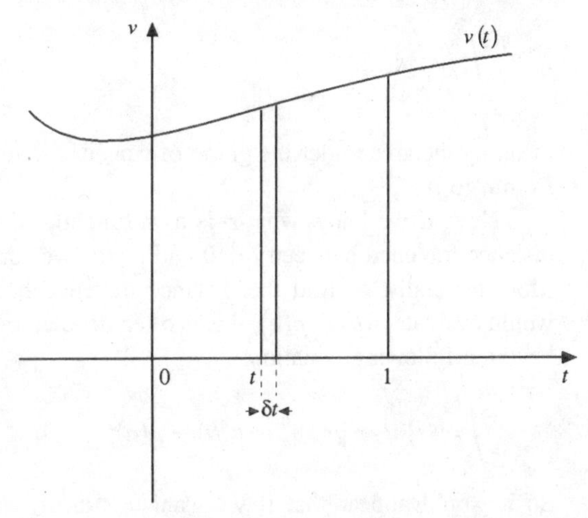

Figure 9.9
Graph of speed against time

So, let us consider the graph of speed against time in Figure 9.9, and see how far the point travels between times $t = 0$ and $t = 1$. In the short time interval from t to $t + \delta t$, the distance travelled is approximately $v(t)\delta t$, and as we see from the diagram, this distance is approximately the area of a rectangle that has height $v(t)$ and width δt. By splitting up the interval into very small sections, we find that the distance travelled is approximated by a large number of such rectangles, as is shown in Figure 9.10, and it is at least plausible that

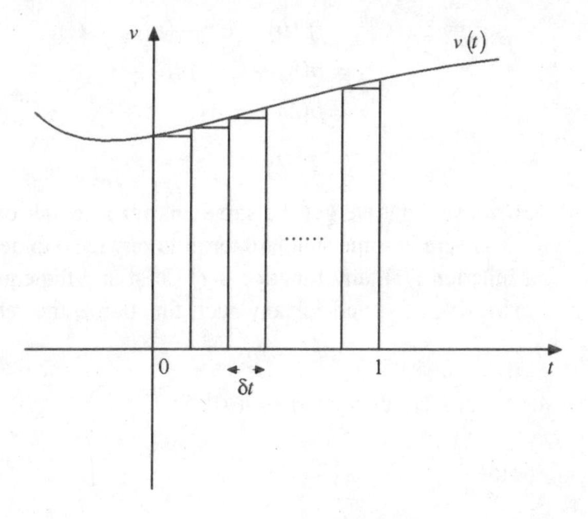

Figure 9.10
Distance travelled approximated by a large number of rectangles

the total distance travelled is just the area under the graph of $v(t)$ between $t = 0$ and $t = 1$. In fact, a more detailed analysis shows that this is exactly the case; the distance travelled between $t = 0$ and $t = 1$ is in fact the area under the graph of v against t between these two values of t. This quantity is denoted in the following way by mathematicians:

$$\int_0^1 v(t)\mathrm{d}t$$

and is called the **integral** of v from 0 to 1. Similarly, the distance travelled between any two times, say $t = a$ and $t = b$, is denoted by

$$\int_a^b v(t)\mathrm{d}t,$$

meaning the area under the graph of v against t, and is called the integral of v from a to b.

Now, if we know what p is as a function of time, it is easy to find the distance travelled between $t = 0$ and $t = 1$; we simply work out $p(1) - p(0)$. More generally, to find the distance travelled between $t = a$ and $t = b$ we would evaluate $p(b) - p(a)$; this is often denoted by $[p(t)]_a^b$. In other words, we have the following equation:

$$\int_a^b v(t)\mathrm{d}t = [p(t)]_a^b = p(b) - p(a)$$

So it would appear that if we want to find $\int_a^b v(t)\mathrm{d}t$, all we have to do is find the function p which gives v when we differentiate it, and then find $p(b) - p(a)$. There is a slight problem here, in that there is more than one such function; since the derivative of a constant is zero we could add any constant to any function whose derivative was v and get another such function. Fortunately, this does not matter. For consider the two functions $p(t)$ and $p(t) + C$, where C is any constant:

$$\begin{aligned} [p(t) + C]_a^b &= (p(b) + C) - (p(a) + C) \\ &= p(b) + C - p(a) - C \\ &= p(b) - p(a) \\ &= [p(t)]_a^b \end{aligned}$$

So we see that we get the same answer in either case.

There is some standard terminology used to describe this situation. Given a function $v(t)$, any function $p(t)$ satisfying the equation $\mathrm{d}p/\mathrm{d}t = v$ is called a **primitive** of v, and for any such function p the relationship

$$\int_a^b v(t)\mathrm{d}t = p(b) - p(a)$$

holds.

In general, if the derivative of p is v, we write

$$\int v(t)\mathrm{d}t = p(t) + C$$

and call the right-hand side the **indefinite integral** of v. If we want a particular primitive of v, we can choose any convenient value of C (usually zero), since the actual value is immaterial.

Worked example

9.14 What is the indefinite integral of v if $v(t) = 4t^3$?

Solution Recalling the formula for the derivative of t^4 we see that this must be a primitive of v. Thus the indefinite integral is $t^4 + C$.

We are now in a position to use these ideas to find distances, at least when velocity is a sufficiently simple function of time.

Worked example

9.15 If $v(t) = t$, find how far the point moves between $t = 0$ and $t = 1$. Assume that times are in seconds and distances in metres.

Solution We need to find a function whose derivative is v; since we know that the derivative of t^2 is $2t$, it follows that $p(t) = \frac{1}{2}t^2$ will do. Then $p(1) - p(0) = 0.5 - 0 = 0.5$, giving a distance of 0.5 metres.

Alternatively, we could draw the graph and work out the area directly. Figure 9.11 shows the region we are concerned with. The area we are

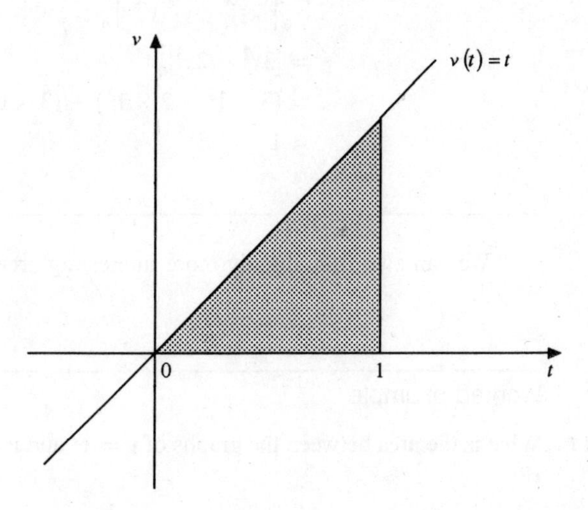

Figure 9.11
Region representing distance travelled by point

interested in is a triangle with base 1 and height 1, and so its area is $\frac{1}{2} \times 1 \times 1 = 0.5$, as before.

Although we have considered the case of position and velocity, the analysis is identical, whatever the dependent and independent variables are called. Thus, if we have the graph $y = f(x)$ and we want to know the area between the graph and the x axis for values of x between a and b, all we need to do is to find some function F whose derivative is f, and then the area is given by $F(b) - F(a)$, which is often written $[F(x)]_a^b$.

First, we write down two rules which follow directly from the rules for derivatives.

KEY POINT

Rule 3 $\int kx^n \mathrm{d}x = \frac{k}{n+1} x^{n+1} + C$ as long as $n \neq -1$.

Rule 4 The integral of the sum of two functions is the sum of their integrals.

Worked example

9.16 What is the area under the graph $y = 6(x - x^2)$ for x between $x = 0$ and $x = 1$?

Solution From the rules above, we have

$$\begin{aligned}
\int_0^1 6(x - x^2)\mathrm{d}x &= \int_0^1 6x - 6x^2 \mathrm{d}x \\
&= \left[6\frac{1}{2}x^2 - 6\frac{1}{3}x^3\right]_0^1 \\
&= [3x^2 - 2x^3]_0^1 \\
&= (3 \times 1^2 - 2 \times 1^3) - (3 \times 0^2 - 2 \times 0^3) \\
&= 1
\end{aligned}$$

We can also find slightly more interesting areas.

Worked example

9.17 What is the area between the graphs of $y = x^2$ and $y = x^3$ for positive values of x?

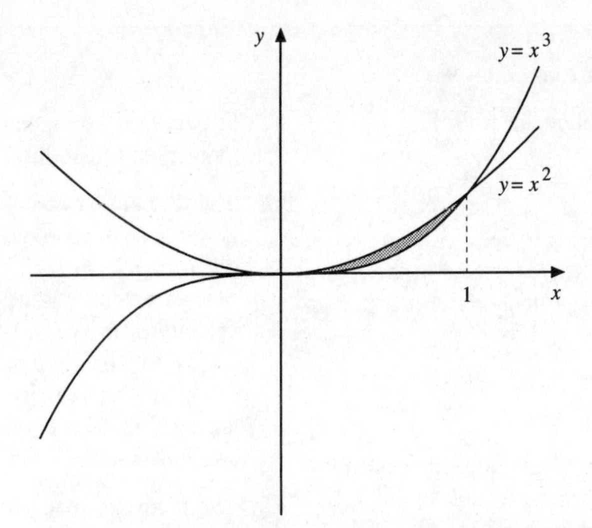

Figure 9.12
Area between the curves $y = x^2$ and $y = x^3$

Solution We begin by drawing the graph of the situation, as shown in Figure 9.12. Then we see that the area we are interested in is given by that of the region under $y = x^2$ and above $y = x^3$, for the values of x between the two values where the two graphs cross. Now, the graphs cross where $x^2 = x^3$, *i.e.* when $x = 0$ and $x = 1$. The area we want is therefore

$$\int_0^1 x^2 \mathrm{d}x - \int_0^1 x^3 \mathrm{d}x = \int_0^1 (x^2 - x^3)\mathrm{d}x$$

$$= \left[\frac{1}{3}x^3 - \frac{1}{4}x^4\right]_0^1$$

$$= \frac{1}{3} - \frac{1}{4}$$

$$= \frac{1}{12}$$

Self-Assessment Questions 9.6

What do you obtain for the area if you consider the graph of $y = -x$ for x between 0 and 5? Why is this a sensible thing to have when the graph is below the x axis?

Exercise 9.6

1. What is the indefinite integral of v if $v(t) = 2t^2$?
2. Find the distance travelled by a point moving with velocity $v(t) = 3$ between $t = 2$ and $t = 5$, both by finding an indefinite integral for v and by directly evaluating the appropriate area.
3. Find the area under the graph $y = x + x^2$ for x between 0 and 2.
4. Find the area between the graphs of $y = 4x^2$ and $y = x^4$ for positive values of x.

Test and Assignment Exercises 9

1. Differentiate the following from first principles:
(i) $y = mx + c$ (ii) $y = x^{-1}$

2. The number of sales per week, S, of a small computer system is related to the price of the system in pounds sterling, p, by the equation

$$S = \frac{760 + p}{p^2}$$

Find the amount by which sales increase if the price is dropped from £1000 to £900. What is the percentage error in the approximation to this you obtain using differentiation?

3. When applied to a certain software house, the Putnam model says that in order to produce a software package of L lines length in one year, the amount of effort required in man-days is $K = L^3/1000$. Approximately how much extra effort is required to increase the number of lines in the package by 10 if the package is (i) 1000 and (ii) 2000 lines long?

4. Differentiate each of the following with respect to the independent variable:
(i) $s = ut + 0.5gt^2$ where u and g are constants, and t is the independent variable.
(ii) $y = (x + 2)(3x^2 - 2x)$
(iii) $v = \frac{2}{\sqrt{t}} - 2\sqrt{t}$ (iv) $s = \left(\frac{2t - 1}{t}\right)^2$
(v) $u = \frac{2v - 3}{v^3}$

5. The resistance R of a component of computer circuitry varies with temperature t according to the formula $R = 200 + 74t + 0.012t^2$. Find the rate at which the resistance is changing with respect to the temperature when the temperature is 100°.

6. The distance in metres, p, a moving point is to the right of some reference point is given by $p = 4t - t^2$, where t is elapsed time in seconds. What is the average velocity of the point between $t = 0$ and $t = 1$? At what time between 0 and 1 is the instantaneous velocity equal to this average? Draw a graph of p against t, and illustrate the situation on it.

7. The predicted load on a computer system over a 12 hour period is modelled by

$$n = 80 - 12t + 2t^2 - \frac{t^3}{12}$$

where n is the number of users logged in and t is the time in hours. Use differentiation to identify the time when the number of users logged in is a minimum, and determine approximately how many users this will be.

8. A manufacturer wishes to house a new computer system in a box which has a square base. If the capacity of the box is to be 8 litres, calculate the dimensions the box should have to minimize the surface area.

9. Find the indefinite integral for each of
(i) $2x^5$ (ii) $x - \sqrt{x}$ (iii) $\frac{x + x^2}{x^4}$

10. Evaluate each of
(i) $\int_1^4 x - \sqrt{x}dx$ (ii) $\int_9^{25} \frac{1}{\sqrt{x}}dx$
(iii) $\int_1^{10} x + x^3 + x^4dx$

11. Find the area of the entire region enclosed between the graphs $y = x^4$ and $y = x^6$.

12. The average value of a function between $x = a$ and $x = b$ is given by

$$\frac{1}{b-a}\int_a^b f(x)\mathrm{d}x$$

(i) Justify this definition by analogy with average velocity.

(ii) What is the average value of x over the range $x = -1$ to $x = 1$? What is the square of this number?

(iii) What is the average value of x^2 over the range $x = -1$ to $x = 1$? Is it the same as the square of the average value of x?

9.7 Further reading

Silvanus P Thompson, *Calculus Made Easy* (St Martin's Press, 1946)
A classic text on calculus taking a friendly approach that is not dominated by the traditional theorem–definition–proof framework.
A Croft and R Davison, *Foundation Mathematics* (Longman, 1994)
A text that covers basic mathematics including calculus, in the same series as this volume.

Appendix: Some basic trigonometry

A.1 The trigonometric ratios

The trigonometric ratios of sine (sin), cosine (cos) and tangent (tan) may be defined in terms of ratios of the lengths of the sides of a right-angled triangle. In Figure A.1 the right-angled triangle ABC contains the angle α. The side opposite α is AB, the side adjacent to α is BC while the hypotenuse of the triangle is AC. Using this triangle we may now state the following definitions.

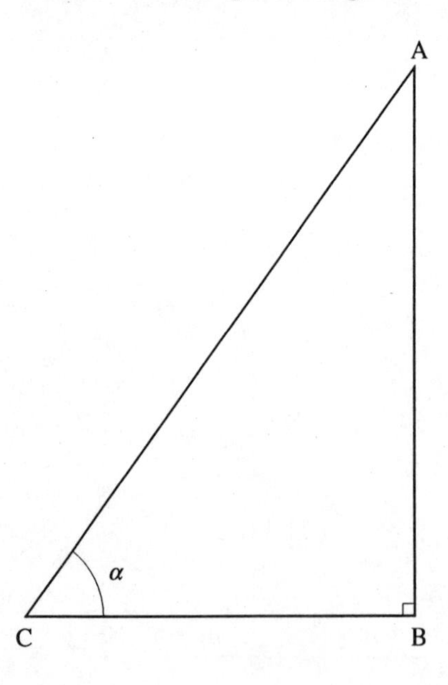

Figure A.1

The sine of an angle is

$$\text{sine of } \alpha = \frac{\text{length of opposite side}}{\text{length of hypotenuse}}$$

and we write

$$\sin \alpha = \frac{AB}{AC}$$

The cosine of an angle is

$$\text{cosine of } \alpha = \frac{\text{length of adjacent side}}{\text{length of hypotenuse}}$$

and we write

$$\cos \alpha = \frac{BC}{AC}$$

The tangent of an angle is

$$\text{tangent of } \alpha = \frac{\text{length of opposite side}}{\text{length of adjacent side}}$$

and we write

$$\tan \alpha = \frac{AB}{BC}$$

A.2 Double angle formulae

From Figure A.2 we see that triangles ABC, DCG and ADE are right angled.

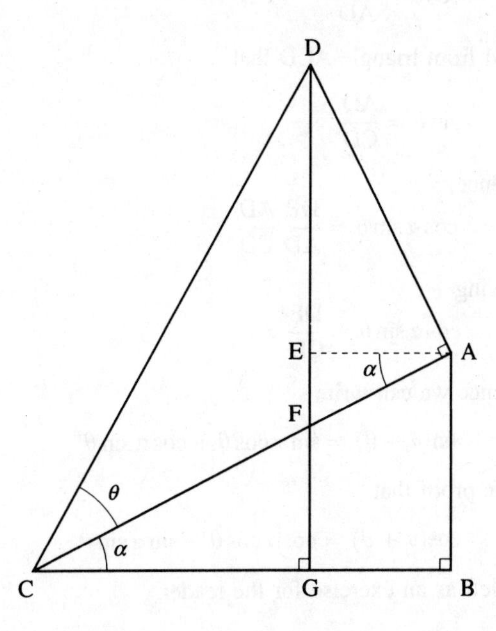

Figure A.2

From triangle DCG we see that

$$\sin(\alpha + \theta) = \frac{DG}{CD}$$

Now DG = DE + EG, hence we may write

$$\sin(\alpha + \theta) = \frac{EG}{CD} + \frac{DE}{CD}$$

From triangle ABC we see that

$$\sin \alpha = \frac{AB}{AC}$$

But AB = EG, hence we may write

$$\sin \alpha = \frac{EG}{AC}$$

From triangle ACD we see that

$$\cos \theta = \frac{AC}{DC}$$

Hence,

$$\sin \alpha \cos \theta = \frac{EG}{AC}\frac{AC}{CD}$$

giving

$$\sin \alpha \cos \theta = \frac{EG}{CD}$$

From triangle ADE we see that

$$\cos \alpha = \frac{DE}{AD}$$

and from triangle ACD that

$$\sin \theta = \frac{AD}{CD}$$

Hence,

$$\cos \alpha \sin \theta = \frac{DE}{AD}\frac{AD}{CD}$$

giving

$$\cos \alpha \sin \theta = \frac{DE}{CD}$$

Hence we can write

$$\sin(\alpha + \theta) = \sin \alpha \cos \theta + \cos \alpha \sin \theta$$

The proof that

$$\cos(\alpha + \theta) = \cos \alpha \cos \theta - \sin \alpha \sin \theta$$

is left as an exercise for the reader.

Solutions to exercises

Solutions to Chapter 1

Exercise 1.1

Compound proposition, not a proposition, compound proposition, simple proposition, not a proposition.

Exercise 1.2

1. My program compiled without any problems $\wedge$ my program had run-time errors.
2. The specification is oversimplified $\vee$ the programming team is unhappy with the specification.
   We can deduce that the specification is oversimplified, since at least one of the simple propositions must be true.
3. No: what the $\Rightarrow$ sign guarantees is that if the first statement is true, then so is the second. It follows that if the second statement is false, the first must be false as well.
4. No: two statements connected by $\Longleftrightarrow$ must both be true or both false.
5. (a) The programming team is unhappy; $\neg$ the programming team is happy. (b) The programming team is sometimes unhappy; $\neg$ the programming team is always happy.
6. (a) The specification is unsuitable and the programming team is happy. (b) If the programming team is unhappy, then the specification is suitable. (c) The specification is suitable or the programming team is unhappy. (d) If the specification is suitable, then the programming team is happy; or the specification is suitable and the programming team is not happy. (e) The specification is suitable if and only if the programming team is happy.
7. (a) $\neg P \vee Q \Rightarrow P \wedge \neg Q$ (b) $P \Rightarrow \neg Q \Longleftrightarrow Q \vee (P \wedge \neg Q)$

Exercise 1.3

1.

P	Q	R	P	$\vee$	$(Q$	$\vee$	$R)$
T	T	T	T	T	T	T	T
T	T	F	T	T	T	T	F
T	F	T	T	T	F	T	T
T	F	F	T	T	F	F	F
F	T	T	F	T	T	T	T
F	T	F	F	T	T	T	F
F	F	T	F	T	F	T	T
F	F	F	F	F	F	F	F
			1	5	2	4	3

P	Q	R	$(P$	$\vee$	$Q)$	$\vee$	R
T	T	T	T	T	T	T	T
T	T	F	T	T	T	T	F
T	F	T	T	T	F	T	T
T	F	F	T	T	F	T	F
F	T	T	F	T	T	T	T
F	T	F	F	T	T	T	F
F	F	T	F	F	F	T	T
F	F	F	F	F	F	F	F
			1	4	2	5	3

Since both final columns are identical, the two propositions are logically equivalent.

2.

P	Q	R	$(P$	$\vee$	$Q)$	$\wedge$	R
T	T	T	T	T	T	T	T
T	T	F	T	T	T	F	F
T	F	T	T	T	F	T	T
T	F	F	T	T	F	F	F
F	T	T	F	T	T	T	T
F	T	F	F	T	T	F	F
F	F	T	F	F	F	F	T
F	F	F	F	F	F	F	F
			1	4	2	5	3

P	Q	R	P	$\vee$	$(Q$	$\wedge$	$R)$
T	T	T	T	T	T	T	T
T	T	F	T	T	T	F	F
T	F	T	T	T	F	F	T
T	F	F	T	T	F	F	F
F	T	T	F	T	T	T	T
F	T	F	F	F	T	F	F
F	F	T	F	F	F	F	T
F	F	F	F	F	F	F	F
			1	5	2	4	3

Since the final columns are different, the two propositions are not logically equivalent, and so the brackets are required.

Exercise 1.4

1. (a)

P	Q	$P \Rightarrow Q$	$\neg P$	$\vee$	Q
T	T	T	F	T	T
T	F	F	F	F	F
F	T	T	T	T	T
F	F	T	T	T	F

Since the column for $P \Rightarrow Q$ is identical to that for $\neg P \vee Q$, the two are logically equivalent.

(b) No; if P is true and Q is false, $P \Rightarrow Q$ is false and $Q \Rightarrow P$ is true.

2. (a) The proposition becomes $\neg(P \wedge Q) \vee P$. This proposition has a truth table

P	Q	$\neg$	$(P$	$\wedge$	$Q)$	$\vee$	P
T	T	F	T	T	T	T	T
T	F	T	T	F	F	T	T
F	T	T	F	F	T	T	F
F	F	T	F	F	F	T	F

and so the proposition is a tautology.
(b) The truth table for $\neg(P \wedge \neg Q)$ is

P	Q	$\neg$	$(P$	$\wedge$	$\neg$	$Q)$
T	T	T	T	F	F	T
T	F	F	T	T	T	F
F	T	T	F	F	F	T
F	F	T	F	F	T	F

which is identical to that for $P \Rightarrow Q$, and so $P \Rightarrow Q \Leftrightarrow \neg(P \wedge \neg Q)$ is a tautology.

Exercise 1.5

1. Use the letter H to represent the simple proposition 'the programming team is happy', and S to represent 'the specification is clear'. Then the argument is $(H \Rightarrow S) \wedge \neg S \Rightarrow \neg H$. We can use algebra to manipulate this as follows:

$$\begin{aligned}(H \Rightarrow S) \wedge \neg S \Rightarrow \neg H &\Longleftrightarrow \neg((\neg H \vee S) \wedge \neg S) \vee \neg H \\ &\Longleftrightarrow ((H \wedge \neg S) \vee S) \vee \neg H \\ &\Longleftrightarrow ((H \vee S) \wedge (S \vee \neg S)) \vee \neg H \\ &\Longleftrightarrow ((H \vee S) \wedge T) \vee \neg H \\ &\Longleftrightarrow H \vee S \vee \neg H \\ &\Longleftrightarrow T\end{aligned}$$

and so the argument is valid.

2. We use J for 'John is on the programming team', A for 'Alice is on the programming team' and R for 'Fred is on the programming team'. The set of statements is inconsistent if $(J \Rightarrow \neg A) \wedge (R \wedge A) \wedge (R \Rightarrow J)$ reduces to F by manipulation.

$$\begin{aligned}&(J \Rightarrow \neg A) \wedge (R \wedge A) \wedge (R \Rightarrow J) \\ \Longleftrightarrow\ &(\neg J \vee \neg A) \wedge (R \wedge A) \wedge (\neg R \vee J) \\ \Longleftrightarrow\ &(\neg J \vee \neg A) \wedge A \wedge (\neg R \vee J) \wedge R \\ \Longleftrightarrow\ &((\neg J \wedge A) \vee (\neg A \wedge A)) \wedge ((\neg R \wedge R) \vee \neg R \wedge J)) \\ \Longleftrightarrow\ &(\neg J \wedge A) \wedge (\neg R \wedge J) \\ \Longleftrightarrow\ &F\end{aligned}$$

so the statements are inconsistent.

Exercise 1.6

1. We have to show that $\neg Q \wedge (P \Rightarrow Q) \Rightarrow \neg P$ is a tautology. Now,

$$\begin{aligned} &\neg Q \wedge (P \Rightarrow Q) \Rightarrow \neg P \\ &\Leftrightarrow \neg(\neg Q \wedge (\neg P \vee Q)) \vee \neg P \\ &\Leftrightarrow (Q \vee (P \wedge \neg Q)) \vee \neg P \\ &\Leftrightarrow (Q \vee P) \vee \neg P \\ &\Leftrightarrow T \end{aligned}$$

which is what we required.

2. First, we pick letters to represent the propositions: (a) D, today is the deadline. (b) S, we must finish the specification. (c) P, we must start the program. (d) U, the team leader is unhappy.
Given this, we can write the hypotheses as (a) $D \Rightarrow S \vee P$ (b) $U \Rightarrow \neg P$ (c) $D \wedge U$
and the conclusion as S. We can now construct a proof.

	Assertion	Reason
1.	$D \wedge U$	Hypothesis 3
2.	D	Hypothesis 3, simplification
3.	$D \Rightarrow S \vee P$	Hypothesis 1
4.	$S \vee P$	2, 3, *modus ponens*
5.	U	1, simplification
6.	$U \Rightarrow \neg P$	Hypothesis 2
7.	$\neg P$	5, 4, *modus ponens*
8.	S	4, 7, disjunctive syllogism

So we see that the conclusion does indeed follow from the hypotheses.

Self-Assessment Questions

1.1 1. A proposition is a sentence that is either true or false. 2. It is a question, and so has no truth value. 3. This is a compound proposition, since it can be split up into two simpler ones connected by the implication connective.

1.2 In each case, we see whether P must be true whenever Q is, or *vice versa*. 1. Whenever P is true, Q must be, so $P \Rightarrow Q$. 2. If P is false, Q must be false. Then if P is true, we don't know anything; however, we cannot have P false and Q true, so $Q \Rightarrow P$. 3. The only time Q is true is when P is true; in other words, P is true whenever Q is, so $Q \Rightarrow P$. 4. Q is a sufficient condition for P; this tells us that whenever Q is true, P is, so $Q \Rightarrow P$. 5. This time, Q is a necessary condition for P; i.e. if P is to be true, Q must be, so $P \Rightarrow Q$.

1.3 No; the construction of the truth table simply provides us with all the information in a structured way, making it easier to see the answer.

1.4 The first statement says that if all the P_i are true, then Q must also be true; the second statement says that it cannot be the case that all the P_i are true and Q is false. These two statements are just different ways of saying the same thing.

1.5 To show that an argument is valid, first express the argument in the form $P_1 \wedge \ldots P_n \Rightarrow Q$, where P_i are the hypotheses and Q is the conclusion. Remove all implication signs by replacing any statement of the form $P \Rightarrow Q$ by one of the form $\neg P \vee Q$. Then use the algebraic laws to simplify the expression. If the argument is valid, it can be reduced to the proposition T.

1.6 Since $P \Longleftrightarrow Q$ is equivalent to $(P \Rightarrow Q) \wedge (Q \Rightarrow P)$, we need to find a proof of $P \vdash Q$ and one of $Q \vdash P$. This tells us that each implies the other, and so the two are logically equivalent.

Solutions to Chapter 2

Exercise 2.1

The elements are 1, $\{1\}$, and $\{1, \{1\}\}$.

Exercise 2.2

1. 5, 6, 7, 8, 9
2. $\{x \in \mathbb{N} | -27 < x^3 < 27\}$
3. (a) $A \cup B = \{1, 3, 4, 5, 6, 7, 8\}$, $A \cup C = \{1, 3, 5, 6, 7, 8, 9, 10\}$, $A \cap B = \{5, 7\}$, $A' = \{2, 4, 6, 8, 9, 10\}$, $C' = \{1, 2, 3, 4\}$, $A \cap (B \cup C) = \{5, 7\}$, $A \cup (B \cap C) = \{1, 3, 5, 6, 7, 8\}$.
   (b) (i) Both are $\{5, 7\}$. (ii) Both are $\{1, 2, 3, 4, 6, 8, 9, 10\}$. (iii) Both are $\{5, 7\}$. (iv) Both are A.
4. (a) (i) $L \cap S$ (ii) $S \cap D$ (iii) $D \cap B$ (iv) $S \cap D \cap B$ (v) $L \cap S \cap B \cap D$.
   (b) (i) Attributes common to sauropods and birds. (ii) Attributes common to lizards and dogs.
5. (a)
$$\begin{aligned} & A \cap (B \cup C) \\ &= \{x \in U | x \in A \wedge (x \in B \vee x \in C)\} \\ &= \{x \in U | (x \in A \wedge x \in B) \vee (x \in A \wedge x \in C)\} \\ &= \{x \in U | (x \in A \cap B) \vee (x \in A \cap C)\} \\ &= (A \cap B) \cup (A \cap C) \end{aligned}$$
   (b) $\{x \in U | T\}$, $\{x \in U | F\}$.

Exercise 2.3

1. We simply construct the Venn diagram for each of these and observe that the two are the same.

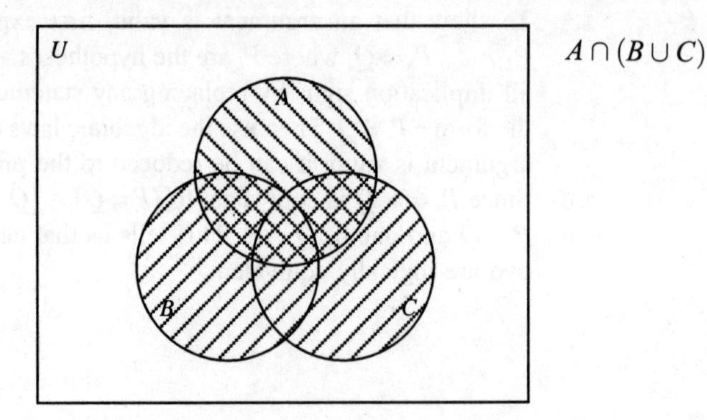

$A \cap (B \cup C)$

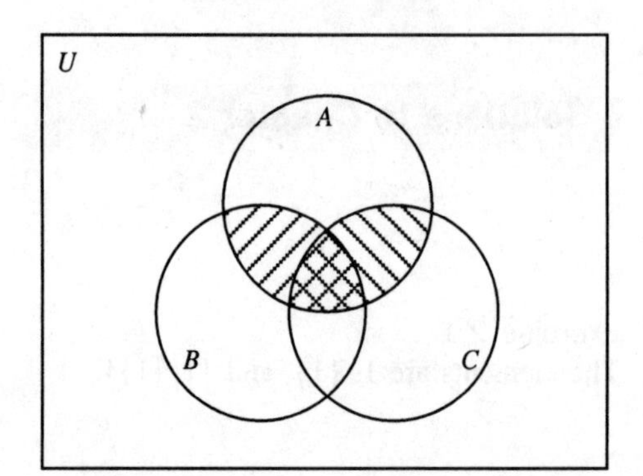

$(A \cap B) \cup (A \cap C)$

2. (a) We construct the membership table for each expression.

A	B	$(A'$	$\cup$	$B')$	$(A \cap B)$	$'$
I	I	O	O	O	I	O
I	O	O	I	I	O	I
O	I	I	I	O	O	I
O	O	I	I	I	O	I

Since the two expressions give the same final column, they define the same set.

(b) Again, we construct a membership table containing each expression.

A	B	C	A	$\cap$	$(B$	$\cap$	$C)$	A	$\cap$	$(B$	$\cup$	$C)$
I	I	I	I	I	I	I	I	I	I	I	I	I
I	I	O	I	O	I	O	O	I	I	I	I	O
I	O	I	I	O	O	O	I	I	I	O	I	I
I	O	O	I	O	O	O	O	I	O	O	O	O
O	I	I	O	O	I	I	I	O	O	I	I	I
O	I	O	O	O	I	O	O	O	O	I	I	O
O	O	I	O	O	O	O	I	O	O	O	I	I
O	O	O	O	O	O	O	O	O	O	O	O	O

Now, we observe that on every row where $A \cap (B \cap C)$ has an I, $A \cap (B \cup C)$ has an I, which is just what we need.

Exercise 2.4

1. As an example, here are the truth tables that check some of the laws of complement.

A	A'	A''	$A \cup A'$	$A \cap A'$	$\emptyset$	U
I	O	I	I	O	O	I
O	I	O	I	O	O	I

2.
$$\begin{aligned} A \cup (A \cap B) &= (A \cap U) \cup (A \cap B) \\ &= A \cap (U \cup B) \\ &= A \cap U \\ &= A \end{aligned}$$

3. (a)
$$\begin{aligned} A \setminus (B \cap C) &= A \cap (B \cap C)' \\ &= A \cap (B' \cup C') \\ &= (A \cap B') \cup (A \cap C') \\ &= (A \setminus B) \cup (A \setminus C) \end{aligned}$$

(b)
$$\begin{aligned} (A \cup B) \setminus (A \cap B) &= (A \cup B) \cap (A \cap B)' \\ &= (A \cup B) \cap (A' \cup B') \\ &= (A \cap (A' \cup B')) \cup (B \cap (A' \cup B')) \\ &= ((A \cap A') \cup (A \cap B')) \cup ((B \cap A') \cup (B \cap B')) \\ &= (\emptyset \cup (A \setminus B)) \cup ((B \setminus A) \cup \emptyset) \\ &= (A \setminus B) \cup (B \setminus A) \end{aligned}$$

Exercise 2.5

1. (a) $12 + 22 = 34$, so 7 people must know both. (b) $5 + 4 = 9$, so 2 can use modules in both.
2. Let's suppose that the teams have been constructed. Then there is some number, say n, of people in all three teams. Then in each of the overlaps there are $6 - n$ members. The total number then must be $3 \times 17 - 3(6 - n) + n = 33 + 4n = 40$, so that $4n = 7$. But this would require fractional people, so it cannot be done.

Exercise 2.6

1. The range of f is $\{2, 4\}$.
2.

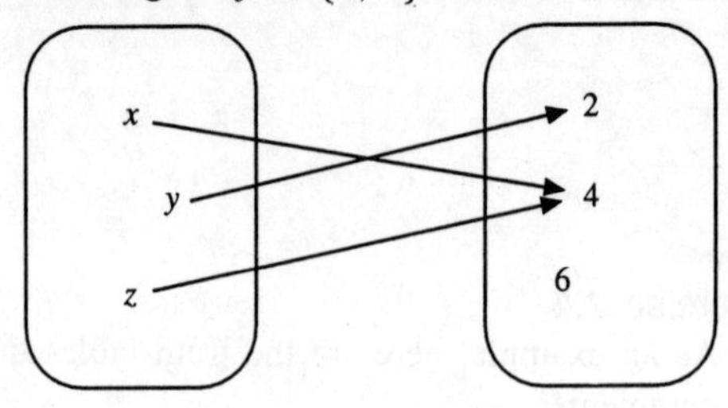

3. (a) None of them.

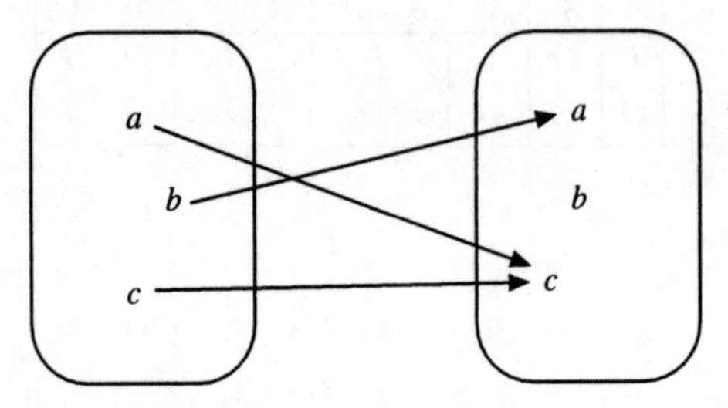

(b) Bijective.

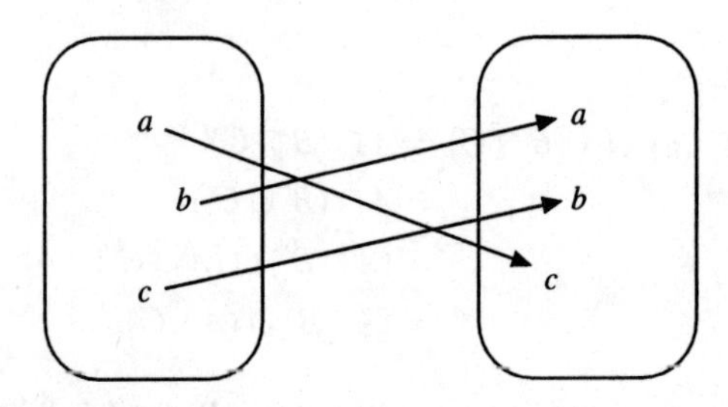

(c) Injective.

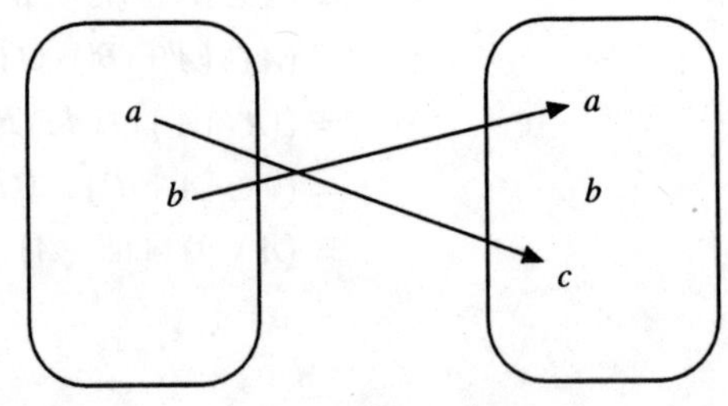

4. $g \circ f(x) = \alpha$, $g \circ f(y) = \omega$, $g \circ f(z) = \alpha$, $g \circ f(w) = \alpha$.

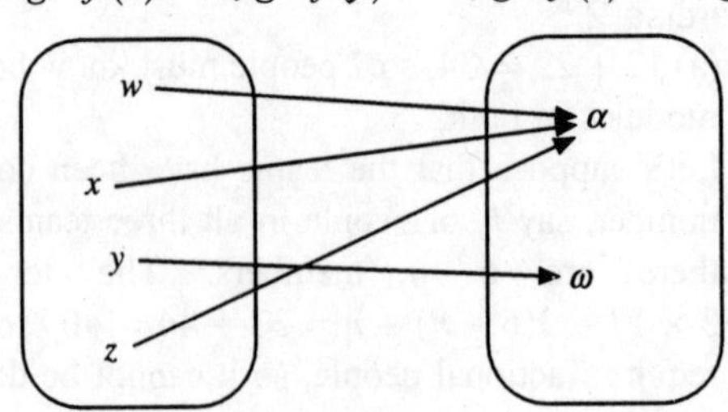

5. (a) Not injective or surjective, therefore no inverse. (b) $f^{-1}(a) = b$, $f^{-1}(b) = c$, $f^{-1}(c) = a$. (c) Not surjective, therefore no inverse.
6. (a) First, we see that $f \circ f$ is not defined if $x = 1$ or if $f(x) = 1$, so the domain of $f \circ f$ is $\mathbb{R} \setminus \{0, 1\}$. On this domain,

$$\begin{aligned} f \circ f(x) &= \frac{1 + (1+x)/(1-x)}{1 - (1+x)/(1-x)} \\ &= \frac{(1-x) + (1+x)}{(1-x) - (1+x)} \\ &= \frac{2}{-2x} \\ &= \frac{1}{-x} \end{aligned}$$

(b) $g \circ f$ has domain $\mathbb{N}$. For any $n \in \mathbb{N}$, $g \circ f(n) = g(f(n)) = g(\sin(n))$ $= \sin^2(n)$. (c) It is not invertible, since $f(1) = f(-1)$. (d) Yes: $f^{-1}(x)$ $= x^{1/3}$.

Exercise 2.7

The set of title–author pairs including FDostoyevsky is p_2^{-1}(FDostoyevsky), so the set of all titles with FDostoyevsky as author is $p_1(p_2^{-1}(\text{FDostoyevsky}))$. Hence the set of all title–author pairs with a title common with one of FDostoyevsky's is $p_1^{-1}(p_1(p_2^{-1}(\text{FDostoyevsky})))$ and finally the required set of author names is $p_2(p_1^{-1}(p_1(p_2^{-1}(\text{FDostoyevsky}))))$.

Self-Assessment Questions

2.1 Without it, one could end up with inconsistent mathematics.

2.2 Yes, because the set-theoretic constructions ∪ and ∩ are defined directly in terms of ∨ and ∧, respectively; arguments about unions and intersections can be phrased with conjunctions and disjunctions replacing the unions and intersections.

2.3 The basic regions in a Venn diagram match up to the row of a membership table in a very direct fashion. For each row, form the expression $A \cap B \cap C$ and take the complement of each set which is labelled by an O rather than an I.

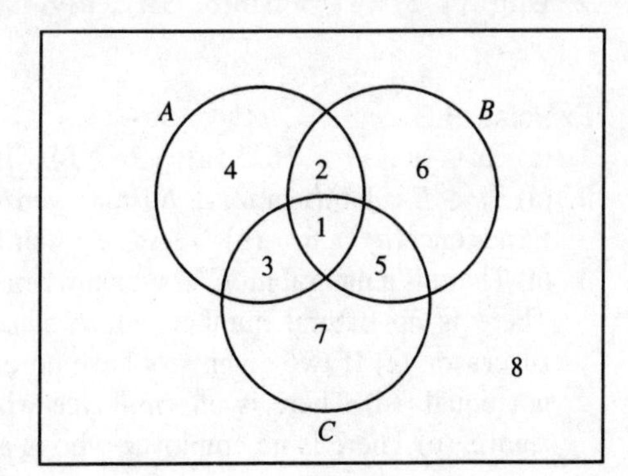

A	B	C	
I	I	I	1
I	I	O	2
I	O	I	3
I	O	O	4
O	I	I	5
O	I	O	6
O	O	I	7
O	O	O	8

This expression corresponds to the region of the Venn diagram consisting of elements in each set labelled with an I, but outside each set labelled with an O.

2.4 Because of the way that union and intersection can be expressed using propositional calculus; alternatively, because of the structural similarity of membership tables and truth tables.

2.5 We would run the risk of counting objects more than once. We can do it if the sets are all disjoint, i.e. if no two sets have any elements in common.

2.6 It means that a program can never give the same output when run on different inputs. The inverse function calculates the input, given the output.

2.7 Functions are inadequate because the relationships are such that one item must be related to many, not just one; a function would only allow (for example) one title per author, or one author per title. It would not allow the possibility of some author writing more than one book and different authors writing books with the same title.

Solutions to Chapter 3

Exercise 3.1

1. (a) Introduce a predicate HadWives(p, n) which takes two variables, p from the set of people and n from the set of natural numbers. We can then write HadWives(HenryVIII, 6). (b) Introduce a predicate Divides(n1, n2), where n1 and n2 are natural numbers and the predicate returns the value *true* when n1 is exactly divisible by n2. We can then write Divides(16, 4). (c) Introduce a predicate IsDenserThan(s1, s2), which returns the value *true* if substance s2 is denser than substance s1. We can then write IsDenserThan(Iron, Lead)
2. (a) $n^2 > 2n$ (b) $xy > x + y$ (c) $\sqrt{n} \leq n$

Exercise 3.2

1. Manages(Paul, Sarah) $\Rightarrow$ Experience(Sarah) $>$ Experience(Paul) $\vee$ (ExpertIn(OOD, Sarah) $\wedge$ ExpertIn(C++, Sarah))
2. $\text{IsIn}(S_1 \cap S_2, e) \Rightarrow \text{IsIn}(S_1 \cup S_2, e) \wedge \text{IsIn}(S_1, e) \wedge \text{IsIn}(S_2, e)$

Exercise 3.3

1. (a) $\exists x \in \mathbb{N} : x^2 > 16$. (b) $\neg \exists x \in \mathbb{N} : \text{IsPrime}(x^2)$.
2. (a) $\exists e \in E :$ IsIn(team2, e) $\wedge$ EmployedAs(programmer, e)$\wedge$ Experience(e) $= 6$. (b) $\neg \exists e \in E :$ IsIn(team1, e) $\wedge$ IsIn(team2, e).
3. (a) There is a natural number which when doubled is equal to its square. (b) There is no natural number whose square is equal to the square of its successor. (c) If two given sets have no common element then the sets are not equal. (d) There is an employee who manages Jill and is leader of team2. (e) There is no employee who is an expert in both COBOL and C.

4. (a) $\forall i \in \mathbb{Z} :$ Divides$(i,\ 2) \Rightarrow$ Even(i). (b) $\forall i \in \mathbb{N} : i^2 \geq 0$. (c) $\forall i,\ j \in \mathbb{Z} :$ Odd$(i) \wedge$ Even$(j) \Rightarrow$ Odd$(i+j)$.
5. (a) If an integer is odd then it is not divisible by 2. (b) The product of any odd and any even integer is even. (c) There is no odd integer whose square is even.
6. (a) $\forall e \in E :$ Expert(C++, e) $\Rightarrow$ EmployedAs(programmer, e).
   (b) $\forall e \in E :$ Leader$(t,\ e) \Rightarrow$ Experience$(e) \geq 3$.
7. (a) $\neg\, \exists i \in \mathbb{Z} : 32 < i < 36 :$ Prime(i).
   (b) $\forall i \in \mathbb{N} : 4 < i < 20 :\ \neg$ Prime$(i) \Rightarrow$ Divides$(i,\ 3) \vee$ Divides$(i,\ 2)$.
   (c) $\exists i \in \mathbb{Z} : 0 \leq i \leq 10 :$ Even$(i) \wedge$ Prime(i).
8. (a) $\forall x \in \mathbb{N} : \neg$ Prime$(x) \Rightarrow \exists y \in \mathbb{N} :$ Prime$(y) \wedge$ Divides$(x,\ y)$.
   (b) $\forall x \in \mathbb{N} : \exists y \in \mathbb{N} : y > x$.
9. (a) $\forall e_1 \in E :$ EmployedAs(systems analyst, e_1) $\Rightarrow$
   $\exists e_2 \in E :$ Manages$(e_1\ ,\ e_2) \wedge$ Experience$(e_2) \geq 5$.
   (b) $\exists t \in PT :\ \forall e \in E :$ IsIn$(t,\ e) \Rightarrow$ ExpertIn(OOD, e).
10. (a) If a natural number is prime then it has no factors other than itself and 1. (b) For every even natural number there is an odd number which is one greater than the given number.

Exercise 3.4

1. (a) $\forall n :\ (n = 2x \Rightarrow \exists k :\ k = 2j + 1 \wedge k = n + 1)$.
   (b) $(\forall n :\ (n = 2i \Rightarrow \exists k :\ (k = 2j + 1 \wedge k = n + 1)))[j\backslash k]$
   $(\forall n :\ (n = 2i \Rightarrow \exists x :\ (x = 2j + 1 \wedge x = n + 1)))[j\backslash k]$
   $\forall n :\ (n = 2i \Rightarrow \exists x :\ (x = 2k + 1 \wedge x = n + 1))$
   (c) $\forall x :\ 0 < x < (2k+1)^2 \text{:}\ (\exists i :\ i = x + 1 \wedge i \leq (2k+1)^2)$
   (d) $\forall x :\ 0 < x < n^2 :\ (\exists i :\ i = x + 1 \wedge i \leq n^2)[n\backslash i]$
   $\forall x :\ 0 < x < n^2 :\ (\exists j :\ j = x + 1 \wedge j \leq n^2)[n\backslash i]$
   $\forall x :\ 0 < x < i^2 :\ (\exists j : j = x + 1 \wedge j \leq i^2)$
2. (a) Manages(Sarah, Jim). (b) Manages(Sarah, Sarah)[Sarah\Jim] Manages(Jim, Jim). (c) $a^2 + 2b + 2ab$.
   (d) $(x^2 + 2y + 2xy)[x\backslash a][a, y\backslash b, a]$
   $(a^2 + 2y + 2ay)[a, y\backslash b, a]$
   $b^2 + 2a + 2ba$

Exercise 3.5

1. First show that $\dfrac{\neg\,(\exists x \in X :\ P(x))}{\forall x \in X :\ \neg\, P(x)}$

Assertion	Reason
1. $\neg\,(\exists x \in X :\ P(x))$	Hypothesis
2. $\forall x \in X :\ \neg\, P(x)$	1. $\forall\exists$ relationship (Rule 10)

Second, show that $\dfrac{\forall x \in X : \neg P(x)}{\neg(\exists x \in X : P(x))}$

Assertion	Reason
1. $\forall x \in X : \neg P(x)$	Hypothesis
2. $\neg(\exists x \in X : P(x))$	1. $\forall\exists$ relationship (Rule 10)

This concludes the justification.

2. (a)

Assertion	Reason
1. $\neg P(x)$	Hypothesis
2. $\forall x \in X : \neg P(x)$	1. $\forall$ introduction (Rule 8)
3. $\neg(\exists x \in X : P(x))$	2. $\forall\exists$ relationship (Rule 10)

This concludes the justification.

(b)

Assertion	Reason
1. $\neg(\exists x \in X : P(x))$	Hypothesis
2. $\forall x \in X : \neg P(x)$	1. $\forall\exists$ relationship (Rule 10)
3. $\neg P(x)[x \backslash a]$	2. $\forall$ remove (Rule 9)

This concludes the justification.

3. First show that $\forall x : P(x) \wedge \forall x : Q(x) \Rightarrow \forall x : (P(x) \wedge Q(x))$.

Assertion	Reason
1. $\forall x : P(x) \wedge \forall x : Q(x)$	Hypothesis
2. $\forall x : P(x)$	1. simplification
3. $P(a)$	2. $\forall$ remove (a arbitrary)
4. $\forall x : Q(x)$	1. simplification
5. $Q(a)$	2. $\forall$ remove (a arbitrary)
6. $P(a) \wedge Q(a)$	3, 5 conjunction
7. $\forall x : (P(x) \wedge Q(x))$	6. $\forall$ introduction

This concludes the first part of the proof.
Second, show that $\forall x : (P(x) \wedge Q(x)) \Rightarrow \forall x : P(x) \wedge \forall x : Q(x)$.

Assertion	Reason
1. $\forall x : P(x) \wedge Q(x)$	Hypothesis
2. $P(a) \wedge Q(a)$	1. $\forall$ remove (a arbitrary)
3. $P(a)$	2. simplification
4. $\forall x : P(x)$	3. $\forall$ introduction
5. $Q(a)$	2. simplification
6. $\forall x : Q(x)$	5. $\forall$ introduction
7. $\forall x : P(x) \wedge \forall x : Q(x)$	4, 6 conjunction

This concludes the proof.

Exercise 3.6

1. The phrase 'any employee who is a team leader should not be a programmer' may be expressed as 'there is no employee who is both a team leader and a programmer'. We may represent this phrase as

$$\neg \exists e \in E: (\exists t \in PT: \text{Leader}(t, e) \wedge \text{EmployedAs}(\text{programmer}, e))$$

The predicate $\forall e \in E: (\exists d \in E: \text{Manages}(d, e) \Rightarrow \exists t \in PT: \text{Leader}(t, e))$ is from Worked example 3.20.

The concluding phrase that 'any employee who is a manager cannot be a programmer' may be expressed as 'there is no employee who is both a manager and a programmer'. We may represent this phrase as

$$\neg \exists e \in E: (\exists d \in E: \text{Leader}(d, e) \wedge \text{EmployedAs}(\text{programmer}, e))$$

We must now show that the conclusion is implied by the hypotheses. This we do as follows.

Assertion	Reason
1. $\neg \exists e \in E: (\exists t \in PT: \text{Leader}(t, e) \wedge \text{EmployedAs}(\text{programmer}, e))$	Hypothesis
2. $\forall e \in E: (\exists d \in E: \text{Manages}(d, e) \Rightarrow \exists t \in PT: \text{Leader}(t, e))$	Hypothesis
3. $\forall e \in E: \neg(\exists t \in PT: \text{Leader}(t, e) \wedge \text{EmployedAs}(\text{programmer}, e))$	1. de Morgan
4. $\forall e \in E: \neg \exists t \in PT: \text{Leader}(t, e) \vee \neg \text{EmployedAs}(\text{programmer}, e)$	3. de Morgan
5. $\forall e \in E: \exists t \in PT: \text{Leader}(t, e) \Rightarrow \neg \text{EmployedAs}(\text{programmer}, e)$	4. $\neg P \vee Q \Leftrightarrow P \Rightarrow Q$
6. $\exists t \in PT: \text{Leader}(t, \text{Sue}) \Rightarrow \neg \text{EmployedAs}(\text{programmer}, \text{Sue})$	5. $\forall$ remove 'Sue' arbitrary
7. $\exists d \in E: \text{Manages}(d, \text{Sue}) \Rightarrow \exists t \in PT: \text{Leader}(t, \text{Sue})$	6. $\forall$ remove 'Sue' arbitrary
8. $\exists d \in E: \text{Manages}(d, \text{Sue}) \Rightarrow \neg \text{EmployedAs}(\text{programmer}, \text{Sue})$	7, 6 hypothetical syllogism
9. $\neg \exists d \in E: \text{Manages}(d, \text{Sue}) \vee \neg \text{EmployedAs}(\text{programmer}, \text{Sue})$	8. $\neg P \vee Q \Leftrightarrow P \Rightarrow Q$
10. $\neg(\exists d \in E: \text{Manages}(d, \text{Sue}) \wedge \text{EmployedAs}(\text{programmer}, \text{Sue}))$	9. de Morgan
11. $\forall e \in E: \neg(\exists d \in E: \text{Manages}(d, e) \wedge \text{EmployedAs}(\text{programmer}, e))$	10. $\forall$ introduction 'Sue' arbitrary
12. $\neg \exists e \in E: (\exists d \in E: \text{Manages}(d, e) \wedge \text{EmployedAs}(\text{programmer}, e))$	11. de Morgan

This concludes the proof.

2. The key phrases 'there is an expert in OOD who manages all analysts' and 'experts on OOD will not manage programmers' may be represented as

$$\exists e \in E: (\text{ExpertIn}(\text{OOD}, e) \wedge$$
$$(\forall d \in E: \text{EmployedAs}(\text{analyst}, d) \Rightarrow \text{Manages}(d, e)))$$

and

$$\forall e, d \in E: \text{ExpertIn}(\text{OOD}, e) \wedge \text{Manages}(d, e) \Rightarrow$$
$$\neg \text{EmployedAs}(\text{programmer}, d)$$

The conclusion that 'no employee is both an analyst and a programmer' may be represented as

$$\neg \exists e \in E: \text{EmployedAs}(\text{analyst}, e) \wedge \text{EmployedAs}(\text{programmer}, e)$$

The justification of the conclusion, using abbreviated predicates, is as follows.

Assertion	Reason
1. $\exists e \in E$: (ExptIn(OOD, e) $\wedge$ ($\forall d \in E$: EmpAs(analyst, d) $\Rightarrow$ Manages(d, e)))	Hypothesis
2. $\forall e, d \in E$: ExptIn(OOD, e) $\wedge$ Manages(d, e) $\Rightarrow \neg$ EmpAs(programmer, d)	Hypothesis
3. ExptIn(OOD, Sue) $\wedge$ (EmpAs(analyst, Joe) $\Rightarrow$ Manages(Joe, Sue))	1. $\exists$ remove 'Sue' fresh, $\forall$ remove 'Joe' arbitrary
4. ExptIn(OOD, Sue) $\wedge$ Manages(Joe, Sue) $\Rightarrow \neg$ EmpAs(programmer, Joe)	2. $\forall$ remove 'Joe' and 'Sue' arbitrary
5. ExptIn(OOD, Sue)	3. simplification
6. EmpAs(analyst, Joe) $\Rightarrow$ Manages(Joe, Sue)	3. simplification
7. $\neg$ (ExptIn(OOD, Sue) $\wedge$ Manages(Joe, Sue)) $\vee \neg$ EmpAs(programmer, Joe)	4. $P \Rightarrow Q \Leftrightarrow \neg P \vee Q$
8. $\neg$ ExptIn(OOD, Sue) $\vee \neg$ Manages(Joe, Sue) $\vee \neg$ EmpAs(programmer, Joe)	7. de Morgan
9. $\neg$ Manages(Joe, Sue) $\vee \neg$ EmpAs(programmer, Joe)	5, 8 Disjunctive Syllogism
10. Manages(Joe, Sue) $\Rightarrow \neg$ EmpAs(programmer, Joe)	9. $P \Rightarrow Q \Leftrightarrow \neg P \vee Q$
11. EmpAs(analyst, Joe) $\Rightarrow \neg$ EmpAs(programmer, Joe)	6, 10 Hypothetical Syllogism
12. $\neg$ EmpAs(analyst, Joe) $\vee \neg$ EmpAs(programmer, Joe)	11. $P \Rightarrow Q \Leftrightarrow \neg P \vee Q$
13. $\neg$ EmpAs(analyst, Joe) $\wedge$ EmpAs(programmer, Joe)	12. de Morgan
14. $\forall e \in E$: $\neg$(EmpAs(analyst, e) $\wedge$ EmpAs(programmer, e))	13. $\forall$ Intro' since 'Joe' arbitrary
15. $\neg \exists e \in E$: (EmpAs(analyst, e) $\wedge$ EmpAs(programmer, e))	14. de Morgan

This concludes the proof.

Self-Assessment Questions

3.1 The symbol $\in$ represents a predicate in that we can write $x \in \mathbb{N}$ or $x \in \mathbb{Z}$. If we then assign the value -1 to the variable x, $x \in \mathbb{N}$ becomes $-1 \in \mathbb{N}$ which is *false* and $x \in \mathbb{Z}$ becomes $-1 \in \mathbb{Z}$ which is *true*.

3.3 In this question we are concerned with those students taking an examination; this specifies a range. Since no student took the examination the range is empty and hence any universally quantified predicate over this range will be *true*. Hence it is true to say that all students who sat the examination passed.

3.4 Since there is no number which is 1 larger than itself the predicate $\exists x : x = x + 1$ is always *false* and hence so is the predicate $(\exists x : x = x + 1) \wedge x = 2x$.
However, the predicate $(\exists z : x = z + 1) \wedge x = 2x$ is *true* when $x = 0$ and $z = -1$.

3.5 If there is no value in the domain of a predicate for which it is *false* then it must be *true* for all values of its domain.

Solutions to Chapter 4

Exercise 4.1

Let us suppose first that the reply is 'no'. If the speaker is a compulsive liar, we know that this must be false, and so Albert is a compulsive liar, so he must be this person. If the speaker is telling the truth, then Albert must be honest, and since at least one of them is a compulsive liar, the speaker must be Albert. Thus if the answer is 'no', Albert is the 'team leader'.

Now, what if the reply is 'yes'? If the speaker is a compulsive liar, then we know that Albert is not, and so he must be the other. On the other hand, if the speaker is honest, Albert is not, and so in this case also Albert must be the other.

Exercise 4.2

2 and 3 are primes as is $2+3$.

Exercise 4.3

1. Yes; the argument is of the form $P \Rightarrow Q$, Q, therefore P, not of the form $P \Rightarrow Q$, $\neg Q$, therefore $\neg P$.
2. We begin by assuming that Butch is newest. We can eliminate this immediately since we are told that either Tom or Jerry is. Next, try Tom. We know that Tom is oldest or Butch is newest, and if Tom is not oldest Butch must be newest. But we had just assumed that Tom was newest, which again provides a contradiction. Finally, try Jerry. Since Butch is not newest, Tom must be oldest, leaving Butch in the middle. This gives us Jerry newest, Butch in the middle, and Tom oldest, which is consistent with all the information; since it is the only possibility consistent with the information, this must be the answer.

Exercise 4.4

1. First, note that if there is one team leader, he requires no phone lines and $1(1-1)/2 = 0$, as required. Next, suppose that for k team leaders we require $k(k-1)/2$ lines. If we include a new team leader, we need k more lines. Now, $k(k-1)/2 + k = (k+1)k/2$, which is just what we get if we replace n by $k+1$ in the formula; thus if the formula is correct when $n = k$ it is correct when $n = k+1$.

   Hence, by induction, the formula is correct for all $n \geq 1$.

2. The first two sums are both

$$1 + \frac{1}{4} + \frac{1}{9} + \frac{1}{16} + \frac{1}{25} + \frac{1}{36}$$

and the third is

$$\frac{1}{2}+\frac{2}{3}+\frac{3}{4}+\frac{4}{5}$$

3. First we observe that the value of E is clearly unchanged by one execution of the loop. Next, suppose that E is unchanged after k executions. If we execute the loop again, E has the same value at the end of that execution as at the start, and so it has the same value as before the first execution. Hence if the value is unchanged by k executions it is unchanged by $k+1$ executions. Thus the value of E is unchanged by any number of executions.
4. Let us denote by x, y the values at the beginning of the loop, and by x' and y' the values after a single execution. Then $x'2^{y'} = 2x2^{y-1} = x2^y$. Thus the value of $x2^y$ is unchanged by execution of the loop. When the program starts, $x = 1$ and $y = n$, so the initial value of $x2^y$ is 2^n. When the loop terminates, $y = 0$, so $x2^y = x$. But this must be 2^n, so upon termination, the value in x is 2^n.

Self-Assessment Questions

4.1 The kind of proof we consider here is a sketch, or outline of a formal proof. In principle, one could fill out such a proof to provide a formal proof such as those seen in Chapters 1 and 3.

4.2 I have not shown that there are no even numbers greater than 100 for which the square fails to be a multiple of 4.

4.3 No it would not; I might just not be clever enough to obtain the contradiction.

4.4 The criticism is mistaken because one does not actually assume that $P(k)$ is true for each value of k; what one does is show that for each k, $P(k) \Rightarrow P(k+1)$. The assumption of $P(k)$ is local to the proof of $P(k+1)$ using that assumption – we don't assume $P(k+1)$ in this proof.

Solutions to Chapter 5

Exercise 5.1

1. $A^2 = abcabc$, $ABA = abcaabbbabc$, $\#(A) = 3$, $\#(B) = 5$, $\#(A^2) = 6 = 2\#(A)$, $\#(ABA) = 11 = 2\#(A) + \#(B)$.
2. *aaabbaa*, *abab*, *aabb*.
3. $(a^3b^2)^2a^4b^2$, $(ab)^5$.

Exercise 5.2

1. (a) a and b are regular expressions, therefore so are ab and ba, therefore ab and aba are, so $(ab \vee aba)$ is and finally $(ab \vee aba)^*$ is. (b) Since a and b are regular expressions, so is ab and hence $(ab)^*$, thus so are $(a \vee (ab)^*)$ and $(a \vee (ab)^*)^*$. (c) a and b are regular expressions, so ab and ba are, so $(ab \vee ba)$ is and so is $(ab \vee ba)^*$.
2. (a) $a^*(bb)^*(c(c^2)^* \vee a^*)$ (b) $1(0 \vee 1)^*000$

Exercise 5.3

1. The automaton is given by $S = \{s_0, s_1\}$, $A = \{s_0\}$ and $d(s_0, b) = s_1$, $d(s_0, a) = s_0$, $d(s_1, b) = s_0$, $d(s_1, a) = s_1$.

   (a) On the string *ababab* we obtain

$$\underline{a}babab : s_0$$
$$a\underline{b}abab : s_0$$
$$ab\underline{a}bab : s_1$$
$$aba\underline{b}ab : s_1$$
$$abab\underline{a}b : s_0$$
$$ababa\underline{b} : s_0$$
$$ababab : s_1$$

   So this string is not accepted.

   (b) On the string *bababba* we have

$$\underline{b}ababba : s_0$$
$$b\underline{a}babba : s_1$$
$$ba\underline{b}abba : s_1$$
$$bab\underline{a}bba : s_0$$
$$baba\underline{b}ba : s_0$$
$$babab\underline{b}a : s_1$$
$$bababb\underline{a} : s_0$$
$$bababba : s_0$$

   And so this string is accepted.

2.

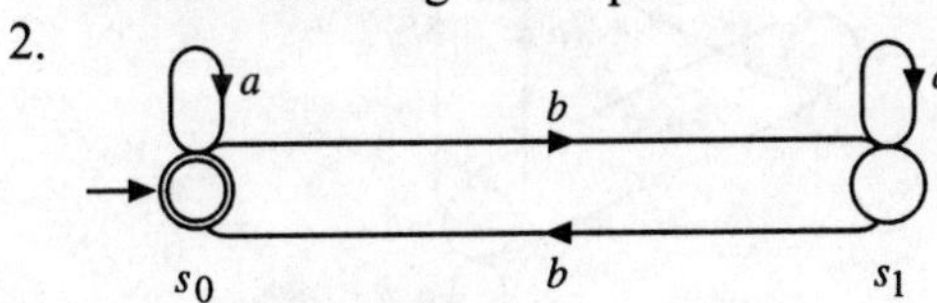

3.

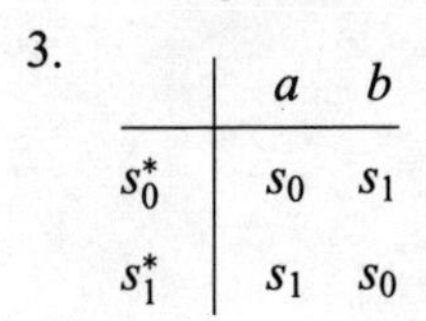

	a	b
s_0^*	s_0	s_1
s_1^*	s_1	s_0

4. (a)

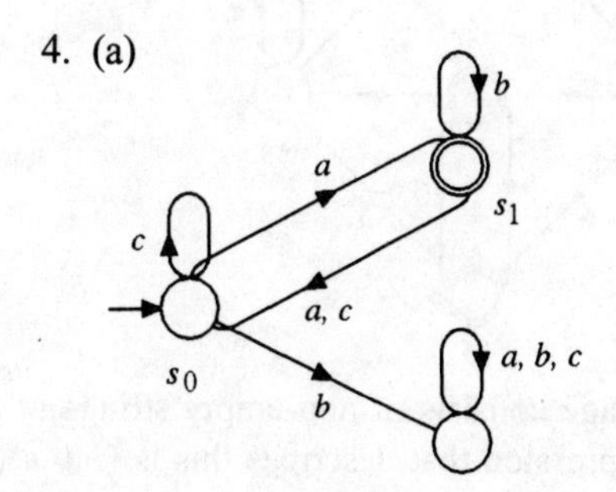

(b)

	a	b	c
s_0	s_1	s_1	s_2
s_1^*	s_3	s_3	s_3
s_2	s_3	s_1	s_3
s_3	s_3	s_3	s_3

(c)

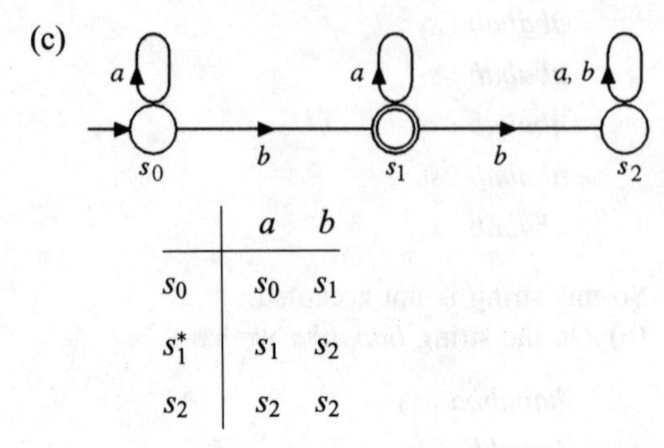

	a	b
s_0	s_0	s_1
s_1^*	s_1	s_2
s_2	s_2	s_2

The strings *baa*, *ab* and *aaba* are accepted, but *aababa* is not.

Exercise 5.4

1. The first and third strings are accepted. The machine recognizes the language given by the regular expression $(01)^*$.

2.

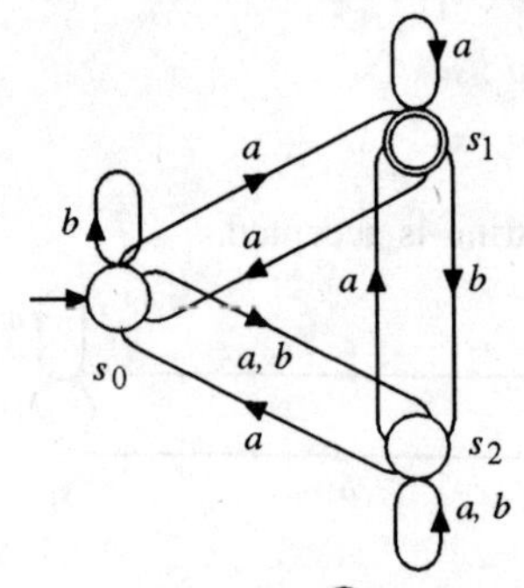

3.

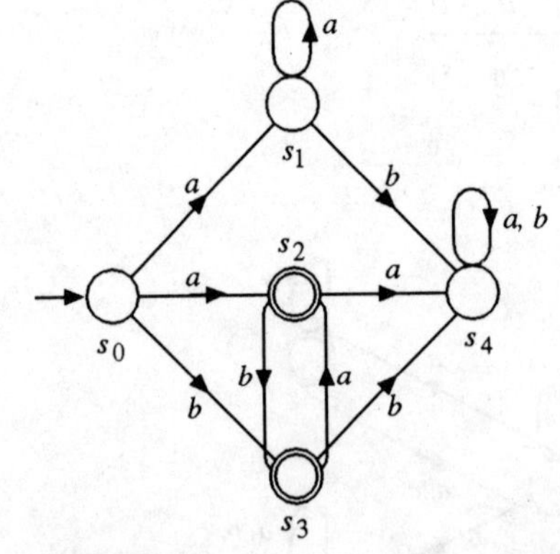

The language consists of non-empty strings of alternating *a*s and *b*s. One regular expression that describes this is $(b \vee \varepsilon)(ab)^*(a \vee \varepsilon) \vee (a \vee b)$.

4. Defining the states S_0 to S_6 by $S_0 = \{s_0\}$, $S_1 = \{s_1\}$, $S_2 = \{s_2\}$, $S_3 = \{s_0, s_1\}$, $S_4 = \{s_0, s_2\}$, $S_5 = \{s_1, s_2\}$, $S_6 = \{s_0, s_1, s_2\}$, we obtain the transition table

	a	b
S_0	S_5	S_4
S_1^*	S_3	S_2
S_2	S_6	S_1
S_3^*	S_6	S_4
S_4	S_6	S_6
S_5^*	S_6	S_5
S_6^*	S_6	S_6

5. L_1

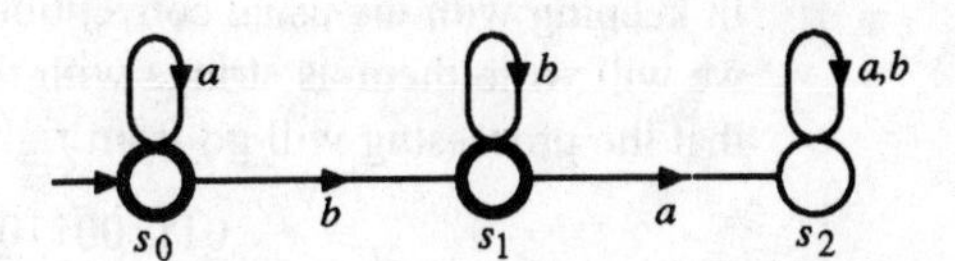

L_2

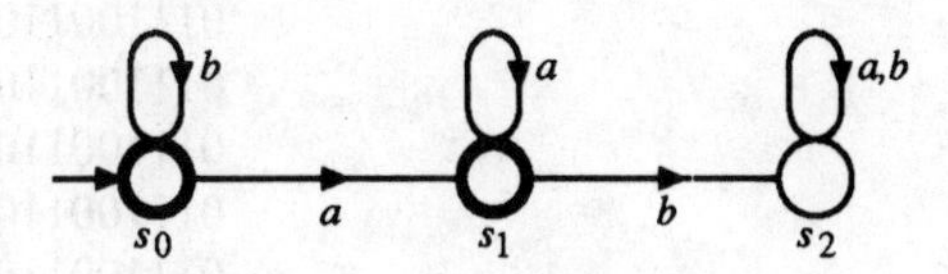

(a) L_1L_2

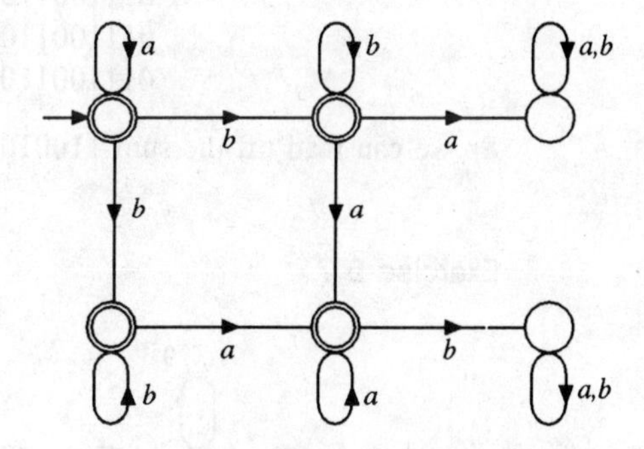

(b) $L_1 \cup L_2$

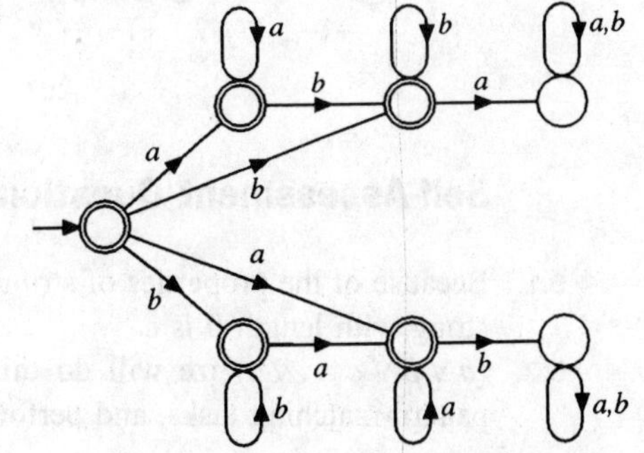

Exercise 5.5

1.

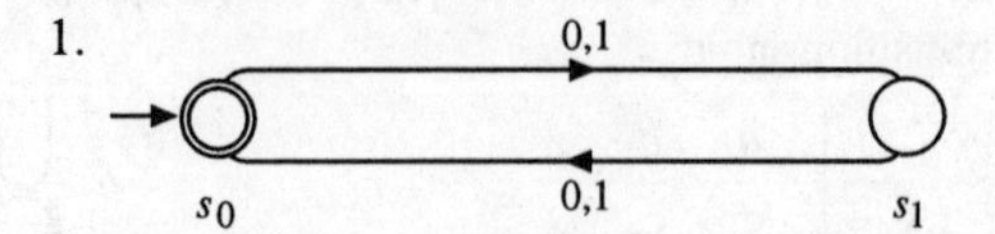

2. Suppose that the language is indeed regular, and is recognized by the machine M which has N states. Consider the string $a^{N+1}b^{N+2}$. While the as are processed, the machine must visit some state twice, so there is a loop of length (say) p, where $p \geq 1$. We can go round this loop as often as we like, so the machine must also recognize $a^{N+p+1}b^{N+2}$, which is not in the language. This contradiction shows that the language is not regular.

Exercise 5.6

In keeping with the usual convention for how binary numbers are represented, we will write them as strings with the units digit to the far right. This means that the processing will go from right to left rather than the usual left to right.

$01110011\underline{0},\ 01011000\underline{1}\quad s_0:$
$0111001\underline{1}0,\ 0101100\underline{0}1\quad s_0:\ 1$
$011100\underline{1}10,\ 010110\underline{0}01\quad s_0:\ 1$
$01110\underline{0}110,\ 01011\underline{0}001\quad s_0:\ 1$
$0111\underline{0}0110,\ 0101\underline{1}0001\quad s_0:\ 0$
$011\underline{1}00110,\ 010\underline{1}10001\quad s_0:\ 1$
$01\underline{1}100110,\ 01\underline{0}110001\quad s_1:\ 0$
$0\underline{1}1100110,\ 0\underline{1}0110001\quad s_1:\ 0$
$\underline{0}11100110,\ \underline{0}10110001\quad s_1:\ 1$
$011100110,\ 010110001\quad s_0:\ 1$

So we can read off the sum 110010111.

Exercise 5.7

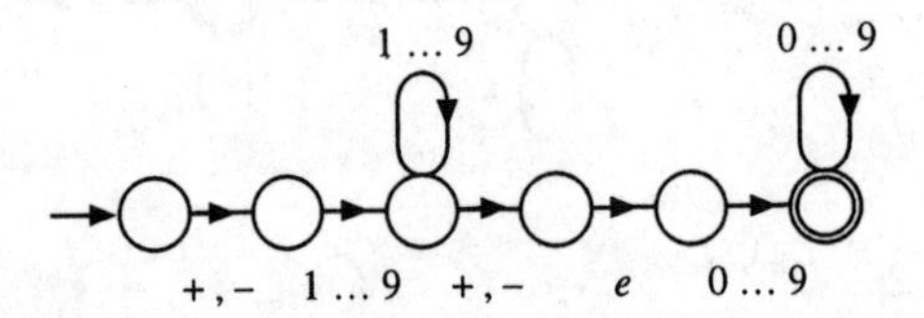

Self-Assessment Questions

5.1 Because of the properties of string length, A must have length 0 and the only string with length 0 is ε.

5.2 $(a \vee b \vee c \ldots \vee z)^*$ize will do this. A program could use this to carry out pattern-matching tasks, and perform (for example) global edits.

5.3 The different representations are useful for different tasks; the tabular method is particularly suitable for writing programs that mimic the behaviour of FSAs, the diagram is easiest for humans to read and understand, the definition in terms of states and transition functions is most useful for consideration of the general properties of automata.

5.4 It is often easier to construct an NDFSA for a given task than to construct a DFSA. Thus, if we want to know whether an FSA can do a job, it is sufficient to construct an NDFSA that does it. Furthermore, if we require a DFSA, we can always construct an NDFSA and then construct an equivalent DFSA from it.

5.5 We can 'talk' to an editor by giving it regular expressions describing patterns we wish to change; the editor can contain code that emulates the behaviour of an FSA that recognizes the regular expression and hence carry out the pattern matching task.

5.6 No, because arithmetic expressions must all have matched brackets, and one particular subset of matched brackets consists of any number of open brackets followed by the same number of closed brackets. We already know that any machine which recognizes matched sets of this form must also recognize unmatched sets, so an FSA that recognizes only legal arithmetic expressions is impossible.

Solutions to Chapter 6

Exercise 6.1

1. You have to do this for yourself.
2. There are 36 possible ways: 11, 12, 13, 14, 15, 16, 21, 22, 23, 24, 25, 26, 31, 32, 33, 34, 35, 36, 41, 42, 43, 44, 45, 46, 51, 52, 53, 54, 55, 56, 61, 62, 63, 64, 65, 66.

   Of these ways, 15, 24, 33, 42 and 51 add up to 6, so there are 5 ways in which the sum can be 6; hence the probability is $5/36$.
3. The sample space is { 11, 12, 13, 14, 15, 16, 21, 22, 23, 24, 25, 26, 31, 32, 33, 34, 35, 36, 41, 42, 43, 44, 45, 46, 51, 52, 53, 54, 55, 56, 61, 62, 63, 64, 65, 66 }. The subset corresponding to the event that the sum is 5 is {14, 23, 32, 41}.
4. (a) {11, 22, 33, 44, 55, 66} (b) {26, 35, 44, 53, 62} (c) {36, 45, 46, 54, 55, 56, 63, 64, 65, 66} (d) {11, 12, 13, 14, 15, 16, 23, 24, 25, 31, 32, 33, 34, 41, 42, 43, 51, 52, 61}
5. (a) (i) Throwing a double 4. (ii) Throwing anything but 8. (iii) Throwing anything but 8. (iv) Throwing an 8 by any means other than a double 4. (b) (i) A' (ii) $A \cap C'$ (iii) $A \cap D$
6. The probability of getting 6 is $5/36$. The probability of a double is $6/36$. These events are not mutually exclusive, since a double 3 gives a score of 6. The probability of getting a 6 or a double is therefore $5/36 + 6/36 - 1/36 = 10/36$.

7. The probability of getting a double is $6/36 = 1/6$. The probability of not getting a double is therefore $1 - 1/6 = 5/6$.

Exercise 6.2

1. There are $6! = 720$ ways of listing the first six letters of the alphabet. If we insist that a comes first and f last, we have $4! = 24$ ways, so the probability of getting this at random is $24/720 = 1/30$.
2. There are $4! = 24$ anagrams of *word*.
3. There are $10!/(2!2!2!) = 453600$ anagrams of *repetition*.
4. There are ${}^7P_4 = 840$ ways.
5. There are ${}^{11}P_5 = 55440$ merit lists.
6. I can choose ${}^{30}C_8 = 5852925$ teams.
7. I have ${}^{14}C_7 = 3432$ choices.
8. I can equip the lab in $4^{10} = 1048576$ ways.
9. The number of ways they can buy computers is $5^4 = 625$. The number of ways they can all pick the same model is 5; therefore the probability that they will all pick the same model is $5/625 = 1/125$.
10. 0. There is no chance of choosing 6 different items from a list of 4.
11. The number of ways of picking 4 from 6 with repetition allowed is ${}^9C_4 = 126$. The number of ways of choosing 4 different items from a list of 6 is ${}^6C_4 = 15$, so the probability is $15/126 \approx 0.119$.

Exercise 6.3

1. (a) The probability that they all pick the same computer is $(1/20)^3 = 1/8000$. (b) No; this does not allow for the possibility that two pick the same computer.
2. (a) The probability of picking a PC is $4/9$, so the probability of picking a PC on each occasion is $(4/9)^4$, which is approximately 0.039.
   (b) The probability of using the PC first, and the mainframe the other three times is $4/9 \times (1/3)^3$, or approximately 0.0165. The PC could be used first, second, third or fourth, so the probability of using the PC once and the mainframe three times is $4 \times 4/9 \times (1/3)^3$, or approximately 0.066.
   (c) The probability of not using a workstation on each occasion is $2/9$, so the probability of not using a workstation on any occasion is $(2/9)^4$. Thus the probability of using a workstation at least once is $1 - (2/9)^4$, or approximately 0.9976.

Exercise 6.4

1. Given that one woman wins first prize and each member of the team is eligible for only one prize, the probability that second prize goes to a woman is $2/9$ (since we have 2 women in our reduced team of 9).
2. Let B denote the event that the listing belonged to Barbara, and C the event that the listing was in C++. Since Alberta brought twice as many listings to the meeting as Barbara, $P(B) = 1/3$. The proportion of C++ listings is $1/4 \times 2/3 + 4/5 \times 1/3 = 13/30$. Hence, $P(C) = 13/30$. Finally, since

four-fifths of Barbara's work is in C++, $P(C|B) = 4/5$. We thus obtain the result that

$$P(B|C) = \frac{P(B)}{P(C)} P(C|B)$$
$$= \frac{1/3}{13/30} \times 4/5$$
$$= \frac{8}{13}$$

3. We need to work out $P(G|n)$. Now,

$$P(G|n) = \frac{P(G)}{P(n)} P(n|G)$$

and $P(G) = 0.98$, $P(n) = 1 - P(p) = 0.788$ and $P(n|G) = 0.8$, so $P(G|n) = 0.98/0.788 \times 0.8$, or approximately 0.995.

Exercise 6.5

1. The probability that the machine was quiet at 8:00 pm is $0.5^3 + 0.5^2 \times 0.4 + 0.5 \times 0.6 \times 0.4 = 0.345$.
2. The probability that the machine was busy all day is 0, since it was quiet at 8:00 am.
3. The probability that the machine was busy on two examinations is $0.5^2 \times 0.6 + 0.5^2 \times 0.4 + 0.5 \times 0.6 \times 0.4 = 0.37$.
4. The probability that the machine was busy on at least two examinations is $0.37 + 0.5 \times 0.6^2 = 0.55$.

Self-Assessment Questions

6.1 If A and B are not mutually exclusive, then the events in $P(A \cap B)$ will be counted twice.

6.2 This is advisable in order to avoid arithmetic overflow when factorials of large integers occur.

6.3 If two events A and B are exclusive, then $P(A \cap B) = 0$. If they are independent, then $P(A \cap B) = P(A)P(B)$. For these two to be equal, at least one of $P(A)$ and $P(B)$ must be zero, i.e. one of the events must be impossible, so this cannot happen for any possible outcomes.

6.4 $P(A|B) = P(B|A)$ will hold when $P(A) = P(B)$, i.e. when A and B are equally likely.

6.5 Probability trees enable us to carry out a careful analysis of all possible cases in a reasonably mechanical way, thereby increasing the probability of carrying out the analysis correctly.

Solutions to Chapter 7

Exercise 7.1

1. Sales achieved in May expressed in £s

	North	South	East	West
Jill	200	400	0	150
Jack	200	300	0	0
Jane	0	310	0	500
John	0	600	0	0

2. (a) 3×3 (b) 3×2 (c) 2×3 (d) 3×4
3. (a) 9 (b) 8.4 (c) $\sqrt{2}$ (d) 6.2

Exercise 7.2

1. (a) $\begin{bmatrix} 4 & 6 & 16 \\ 15 & 1 & 11 \\ 13 & 10 & 10 \end{bmatrix}$ (b) $\begin{bmatrix} 2x+1 \\ y^2+y \\ 4z \end{bmatrix}$

   (c) Matrices $\boldsymbol{F}$ and $\boldsymbol{E}$ have different dimensions and so cannot be added together.

   (d) $\begin{bmatrix} 2x+1 \\ y^2+y \\ 4z \end{bmatrix}$

   (e) $\begin{bmatrix} 2a+\sqrt{2} & 2b+c+6 \\ c+d+4 & d^2+8 \end{bmatrix}$

2. (a) $\begin{bmatrix} 0 & 6 & 8 \\ 13 & -1 & 7 \\ 7 & 6 & 2 \end{bmatrix}$ (b) $\begin{bmatrix} a-\sqrt{2} & b-6 \\ c-4 & d-8 \end{bmatrix}$

   (c) $\boldsymbol{C}-\boldsymbol{J}$ cannot be calculated as the matrices have different dimensions.

   (d) $\begin{bmatrix} -a & -b+c \\ -d & d-d^2 \end{bmatrix}$

   (e) $\boldsymbol{B}-\boldsymbol{M}$ cannot be calculated as the matrices have different dimensions.

Exercise 7.3

1. Target costs $= 90\% \times \begin{bmatrix} 100 & 80 & 20 \\ 90 & 60 & 120 \\ 40 & 50 & 40 \end{bmatrix} = \begin{bmatrix} 90 & 72 & 18 \\ 81 & 54 & 108 \\ 36 & 45 & 36 \end{bmatrix}$

2. $3\boldsymbol{A}+2\boldsymbol{B} = 3\begin{bmatrix} 2 & 3 \\ 6 & 5 \end{bmatrix} + 2\begin{bmatrix} 2 & 4 \\ 5 & 3 \end{bmatrix}$

   $= \begin{bmatrix} 6 & 9 \\ 18 & 15 \end{bmatrix} + \begin{bmatrix} 4 & 8 \\ 10 & 6 \end{bmatrix} = \begin{bmatrix} 10 & 17 \\ 28 & 21 \end{bmatrix} = \boldsymbol{C}$

3. $\begin{bmatrix} 10 & 8 \\ 8 & 14 \end{bmatrix} = \begin{bmatrix} 5 & x \\ 2 & z \end{bmatrix} + 2\begin{bmatrix} y & 3 \\ 3 & 6 \end{bmatrix}$

$= \begin{bmatrix} 5 & x \\ 2 & z \end{bmatrix} + \begin{bmatrix} 2y & 6 \\ 6 & 12 \end{bmatrix} = \begin{bmatrix} 5+2y & 6+x \\ 8 & 12+z \end{bmatrix}$

Hence we require $10 = 5 + 2y \quad 6 + x = 8 \quad 12 + z = 14$

giving: $y = 2\frac{1}{2} \quad x = 2 \quad z = 2$

4. (a) $[13x - 9 \quad 19y - 16]$

(b) $\begin{bmatrix} 3a+2b & 2c+3a \\ 3a+2b & 3c+2a \\ 4a+6b & 4c+6a \end{bmatrix}$

(c) $[2x + 3y + 4z \quad 3x + 2y + 6z]$ (d) $[2x + 3y + 7z]$ (e) ***BG*** cannot be calculated since ***B*** does not have the same number of columns as ***G*** has rows.

5. (a) $\begin{bmatrix} 2x+5y+4z \\ 4x+6y+2z \\ 3x+2y+z \end{bmatrix}$ (b) $\begin{bmatrix} 35 \\ 34 \\ 16 \end{bmatrix}$

(c) ***FH*** cannot be calculated since ***F*** does not have the same number of columns as ***H*** has rows.

(d) $[ax + by + cz]$ (e) $\begin{bmatrix} 2x+y \\ 4x-y \end{bmatrix}$

6. (a) $\begin{bmatrix} 6a+5c & 6b+5d \\ 4a+7c & 4b+7d \\ 2a+3c & 2b+3d \end{bmatrix}$ (b) $\begin{bmatrix} 16 & 21 \\ 28 & 37 \end{bmatrix}$

(c) ***DF*** cannot be calculated as ***D*** does not have the same number of columns as ***F*** has rows.

(d) $\begin{bmatrix} 38 & 54 \\ 38 & 49 \\ 34 & 37 \end{bmatrix}$ (e) $\begin{bmatrix} 8x & 11y \\ 14x & 19y \end{bmatrix}$

(f) C^2 cannot be calculated since the matrix ***C*** is not square, i.e it does not have the same number of rows as columns.

Exercise 7.5

1. (a) $\begin{bmatrix} 5 & 0 & 0 \\ 0 & 1 & 0 \\ 0 & 0 & 1 \end{bmatrix}$ (b) $\begin{bmatrix} 1 & 0 & 0 \\ 0 & d & 0 \\ 0 & 0 & 1 \end{bmatrix}$ (c) $\begin{bmatrix} 1 & 0 & 0 \\ 1 & 1 & 0 \\ 0 & 0 & 1 \end{bmatrix}$ (d) $\begin{bmatrix} 1 & 0 & 0 \\ 0 & 1 & 0 \\ 0 & 1 & 1 \end{bmatrix}$

(e) $\begin{bmatrix} 1 & 0 & 0 & 0 \\ 0 & 1 & 0 & 0 \\ 0 & 0 & 1 & 0 \\ 1 & 0 & 0 & 1 \end{bmatrix}$ (f) $\begin{bmatrix} 0 & 0 & 1 \\ 0 & 1 & 0 \\ 1 & 0 & 0 \end{bmatrix}$ (g) $\begin{bmatrix} 1 & 0 & 0 \\ 0 & 0 & 1 \\ 0 & 1 & 0 \end{bmatrix}$

2. (a) Subtract twice row 1 from row 2. Premultiply by:

$$\begin{bmatrix} 1 & 0 & 0 \\ -2 & 1 & 0 \\ 0 & 0 & 1 \end{bmatrix}$$

(b) Subtract row 1 from row 2 and subtract twice row 1 from row 3. To do this premultiply by

$$\begin{bmatrix} 1 & 0 & 0 \\ -1 & 1 & 0 \\ 0 & 0 & 1 \end{bmatrix} \begin{bmatrix} 1 & 0 & 0 \\ 0 & 1 & 0 \\ -2 & 0 & 1 \end{bmatrix} = \begin{bmatrix} 1 & 0 & 0 \\ -1 & 1 & 0 \\ -2 & 0 & 1 \end{bmatrix}$$

(c) Subtract three times row 1 from row 3 and multiply row 2 by 1/3.

$$\begin{bmatrix} 1 & 0 & 0 \\ 0 & 1 & 0 \\ -3 & 0 & 1 \end{bmatrix} \begin{bmatrix} 1 & 0 & 0 \\ 0 & 1/3 & 0 \\ 0 & 0 & 1 \end{bmatrix} = \begin{bmatrix} 1 & 0 & 0 \\ 0 & 1/3 & 0 \\ -3 & 0 & 1 \end{bmatrix}$$

Exercise 7.6

1. (a) The matrix is not square. (b) Column 3 of the matrix consists only of zeros. (c) Row 1 of the matrix can be generated by subtracting row 3 from row 2. (d) Row 3 of the matrix can be generated by multiplying row 1 by 3.
2. (a) $\begin{bmatrix} 1/5 & 3/5 \\ 1/5 & -2/5 \end{bmatrix}$ (b) $\begin{bmatrix} -2 & 7/6 \\ 1 & -1/2 \end{bmatrix}$ (c) $\begin{bmatrix} 3/11 & -1/11 \\ -1/11 & -2/11 \end{bmatrix}$
3. (a) $\begin{bmatrix} 2/7 & -3/7 \\ 1/7 & 2/7 \end{bmatrix}$ (b) $\begin{bmatrix} -2 & 8 & -19 \\ 1 & -4 & 10 \\ 1 & -3 & 7 \end{bmatrix}$ (c) $\begin{bmatrix} 3/7 & 2/7 & 1/7 \\ -1/7 & 4/7 & 2/7 \\ 1/7 & 3/7 & 5/7 \end{bmatrix}$

Exercise 7.7

1. (a) $x = 3, y = 4$ (b) $x = 1, y = -5, z = -4$, (c) $a = -2, b = 2, c = -1$

Self-Assessment Questions

7.1 A column matrix containing four values would have dimensions 4×1, while a row matrix containing six values would have dimensions 1×6.

7.2 1. Let a_{ij} and b_{ij} be two arbitrary but corresponding elements of the matrices $\boldsymbol{A}$ and $\boldsymbol{B}$. The addition of two matrices is achieved by adding together the corresponding elements from the matrices. Hence the evaluation of $\boldsymbol{A} + \boldsymbol{B}$ would involve the calculation $a_{ij} + b_{ij}$, while the evaluation of $\boldsymbol{B} + \boldsymbol{A}$ would involve the calculation $b_{ij} + a_{ij}$. Since $a_{ij} + b_{ij} = b_{ij} + a_{ij}$ we are justified in stating that generally $\boldsymbol{A} + \boldsymbol{B} = \boldsymbol{B} + \boldsymbol{A}$, i.e. matrix addition is commutative.

2. Using a similar argument for that given for Question 1 we can conclude that it is **not** generally true that $\boldsymbol{A}-\boldsymbol{B} = \boldsymbol{B}-\boldsymbol{A}$ since it is not generally true that $a_{ij} - b_{ij} = b_{ij} - a_{ij}$. Matrix subtraction is not commutative.

3. With arbitrary but corresponding values a_{ij}, b_{ij} and c_{ij} from the matrices $\boldsymbol{A}$, $\boldsymbol{B}$ and $\boldsymbol{C}$, generally we have

$$(a_{ij} + b_{ij}) + c_{ij} = a_{ij} + (b_{ij} + c_{ij})$$

while generally

$$(a_{ij} - b_{ij}) - c_{ij} \neq a_{ij} - (b_{ij} - c_{ij})$$

Hence we are justified in stating that in general

$$(\boldsymbol{A}+\boldsymbol{B})+\boldsymbol{C}=\boldsymbol{A}+(\boldsymbol{B}+\boldsymbol{C})$$

while in general

$$(\boldsymbol{A}-\boldsymbol{B})-\boldsymbol{C}\neq \boldsymbol{A}-(\boldsymbol{B}-\boldsymbol{C})$$

that is to say matrix addition is associative while subtraction is not associative.

7.3 1. Again we base our argument on corresponding arbitrary values a_{ij} and b_{ij} from the matrices $\boldsymbol{A}$ and $\boldsymbol{B}$. The calculation of $n(\boldsymbol{A}+\boldsymbol{B})$ would involve calculating $n(a_{ij}+b_{ij})$, while the calculation of $n\boldsymbol{A}+n\boldsymbol{B}$ would involve calculating $na_{ij}+nb_{ij}$. Since it is generally true that $n(a_{ij}+b_{ij})=na_{ij}+nb_{ij}$ we are justified in stating that it is generally true that $n(\boldsymbol{A}+\boldsymbol{B})=n\boldsymbol{A}+n\boldsymbol{B}$.

2. For the calculations $\boldsymbol{A}(\boldsymbol{B}+\boldsymbol{C})$ and $\boldsymbol{AB}+\boldsymbol{AC}$ to be possible the matrices $\boldsymbol{B}$ and $\boldsymbol{C}$ must have the same dimensions, and what is more they must have the same number of rows as the matrix $\boldsymbol{A}$ has columns. Given that these conditions hold let us consider the calculation $\boldsymbol{AB}+\boldsymbol{AC}$ where $\boldsymbol{A}$ is an $m\times n$ matrix and $\boldsymbol{B}$ and $\boldsymbol{C}$ are $n\times k$ matrices. Let the result of calculating the product be some matrix $\boldsymbol{R}$. If r_{ij} is some arbitrary element of $\boldsymbol{R}$ then we know that

$$r_{ij}=a_{i1}\times b_{1j}+a_{i2}\times b_{2j}+\ \cdots\ +a_{in}\times b_{nj}$$

Similarly the corresponding element of the matrix which results from the calculation of the product $\boldsymbol{AC}$ will have the form

$$a_{i1}\times c_{1j}+a_{i2}\times c_{2j}+\ \cdots\ +a_{in}\times c_{nj}$$

Adding these two results together we will get an expression for the corresponding element in the matrix $\boldsymbol{AB}+\boldsymbol{AC}$. We get

$$\begin{aligned}&\left(a_{i1}\times b_{1j}+a_{i2}\times b_{2j}+\ \cdots\ +a_{in}\times b_{nj}\right)\\&\quad+\left(a_{i1}\times c_{1j}+a_{i2}\times c_{2j}+\ \cdots\ +a_{in}\times c_{nj}\right)\end{aligned}$$

This expression may be written as

$$a_{i1}\left(b_{1j}+c_{1j}\right)+a_{i2}\left(b_{2j}+c_{2j}\right)+\ \cdots\ +a_{in}\left(b_{nj}+c_{nj}\right)$$

It is not difficult to see that this is the value of the element in the ith row and jth column of the matrix that results from the calculation of $\boldsymbol{A}(\boldsymbol{B}+\boldsymbol{C})$. Hence it is reasonable to assume that

$$\boldsymbol{AB}+\boldsymbol{AC}=\boldsymbol{A}(\boldsymbol{B}+\boldsymbol{C})$$

7.4 For the calculation of $\boldsymbol{IA}$ to be possible we require that columns in $\boldsymbol{I}=$ rows in $\boldsymbol{A}$. But for the equation $\boldsymbol{IA}=\boldsymbol{A}$ to hold we require that rows in $\boldsymbol{I}=$ rows in $\boldsymbol{A}$. Since both of these conditions must hold we get columns in $\boldsymbol{I}=$ rows in $\boldsymbol{A}=$ rows in $\boldsymbol{I}$. Assume $\boldsymbol{I}$ is an $n\times n$ matrix. We also require that the calculation $\boldsymbol{AI}$ be possible, which requires that columns in $\boldsymbol{A}=$ rows in $\boldsymbol{I}$.
So $\boldsymbol{A}$ must have n columns and so $\boldsymbol{A}$ is also an $n\times n$ matrix.

Solutions to Chapter 8

Exercise 8.1

1. Point P will have co-ordinate (−6.0), point Q will have co-ordinate (2.1), point R will have co-ordinate (−1.2).
2. For the point P we have

$$-2.5 = 8 + T_P$$
$$T_P = -10.5$$

For the point Q we have

$$8 = -2.5 + T_Q$$
$$T_Q = 10.5$$

3. (a) 3 (b) 12 (c) 11

Exercise 8.2

1.

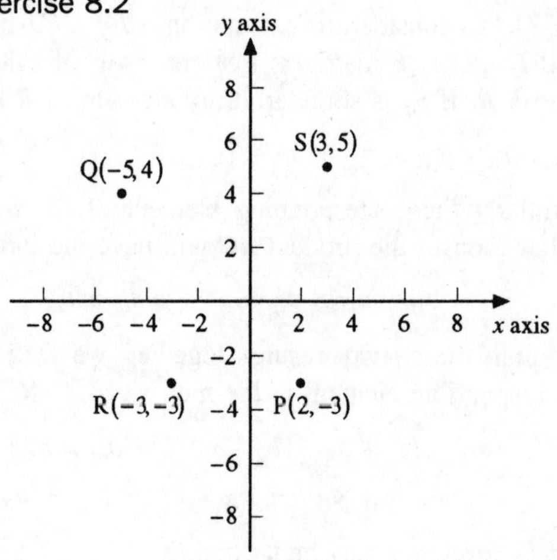

2. Distance $= \sqrt{(10-(-2))^2+(7-2)^2} = \sqrt{144+25} = \sqrt{169} = 13$
3. Length of side AB $= \sqrt{(-7-3)^2+(5-(-1))^2} = \sqrt{100+36} = \sqrt{136}$. Hence $AB^2 = 136$.
Length of side AC $= \sqrt{(-7-1)^2+(5-7)^2} = \sqrt{64+4} = \sqrt{68}$. Hence $AC^2 = 68$.
Length of side BC $= \sqrt{(3-1)^2+(-1-7)^2} = \sqrt{4+64} = \sqrt{68}$. Hence $BC^2 = 68$.
We see that $AB^2 = AC^2 + BC^2$, hence the triangle ABC is right angled.

Exercise 8.3

1. Co-ordinates of P relative to O* are (3, 15), co-ordinates of Q relative to O* are (8, 2).
2. Let the co-ordinates of P and Q relative to O be (x_P, y_P) and (x_Q, y_Q) respectively. Let the co-ordinates of these points relative to O* be (x_P^*, y_P^*)

and (x_Q^*, y_Q^*). Let the co-ordinates of O* be (T_x, T_y) relative to O, then we have

$$x_P^* = x_P - T_x \quad \text{and} \quad x_Q^* = x_Q - T_x$$
$$y_P^* = y_P - T_y \qquad\qquad y_Q^* = y_Q - T_y$$

Relative to O* the distance between P and Q is given by

$$\text{distance*PQ} = \sqrt{(x_P^* - x_Q^*)^2 + (y_P^* - y_Q^*)^2}$$

We may write this as

$$\text{distance*PQ} = \sqrt{((x_P - T_x) - (x_Q - T_x))^2 + ((y_P - T_y) - (y_Q - T_y))^2}$$

which simplifies to distance $*\text{PQ} = \sqrt{(x_P - x_Q)^2 + (y_P - y_Q)^2}$

The expression on the right-hand side of this equation is the distance between P and Q relative to O. Hence we see that distances are not affected by a translation of the origin.

3. The transformation matrix is

$$\begin{bmatrix} 1 & 0 & 0 \\ 0 & 1 & 0 \\ -5 & 8 & 1 \end{bmatrix}$$

The new co-ordinates are (−3, 13), (−8, 15) and (0, 0).

4. The transformation matrix is

$$\begin{bmatrix} 1 & 0 & 0 \\ 0 & 1 & 0 \\ -2 & 4 & 1 \end{bmatrix}$$

The new co-ordinates are (2, 10), (−4, 12) and (−8, 6).

5. The transformation matrix is

$$\begin{bmatrix} \cos 45^\circ & \sin 45^\circ & 0 \\ -\sin 45^\circ & \cos 45^\circ & 0 \\ 0 & 0 & 1 \end{bmatrix} \text{ which is equivalent to } \begin{bmatrix} \sqrt{2}/2 & \sqrt{2}/2 & 0 \\ -\sqrt{2}/2 & \sqrt{2}/2 & 0 \\ 0 & 0 & 1 \end{bmatrix}$$

The vertices of the triangle have the co-ordinates $(-\sqrt{2}, 5\sqrt{2})$, $(-5\sqrt{2}, 3\sqrt{2})$ and $(-4\sqrt{2}, -2\sqrt{2})$. In decimal form these co-ordinates are approximately (−1.41, 7.07), (−7.07, 4.24) and (−5.66, −2.83).

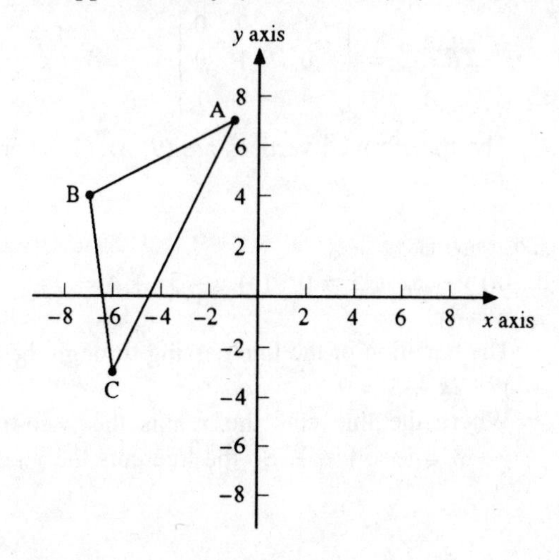

6. Distance between the points (−2, −4), (2, 6) is given by

$$\sqrt{(-2-2)^2+(-4-6)^2} = \sqrt{16+100} = \sqrt{116}$$

The transformation matrix is

$$\begin{bmatrix} \cos 60° & \sin 60° & 0 \\ -\sin 60° & \cos 60° & 0 \\ 0 & 0 & 1 \end{bmatrix} \text{ which is equivalent to } \begin{bmatrix} 1/2 & \sqrt{3}/2 & 0 \\ -\sqrt{3}/2 & 1/2 & 0 \\ 0 & 0 & 1 \end{bmatrix}$$

Transformed co-ordinates are $(2\sqrt{3}-1,\ -2-\sqrt{3})$ and $(1-3\sqrt{3},\ 3+\sqrt{3})$. Distance between these points is

$$\sqrt{\left(\left(2\sqrt{3}-1\right)-\left(1-3\sqrt{3}\right)\right)^2+\left(\left(-2-\sqrt{3}\right)-\left(3+\sqrt{3}\right)\right)^2}$$

$$= \sqrt{\left(5\sqrt{3}-2\right)^2+\left(-5-2\sqrt{3}\right)^2}$$

$$= \sqrt{75-20\sqrt{3}+4+25+20\sqrt{3}+12} = \sqrt{116}$$

So we see that the distance is unaffected by the transformation.

7.

$$[x^*,\ y^*,\ 1] = [6,\ 6,\ 1] \begin{bmatrix} \sqrt{3}/2 & 1/2 & 0 \\ -1/2 & \sqrt{3}/2 & 0 \\ -\sqrt{3}+6 & 7-4\sqrt{3} & 1 \end{bmatrix}$$

$$= \left[2\sqrt{3}+2,\ 10-\sqrt{3},\ 1\right]$$

The co-ordinate is $(2\sqrt{3}+3,\ 10-\sqrt{3})$ which is consistent with the earlier calculation.

8.

$$\boldsymbol{T} = \begin{bmatrix} 1 & 0 & 0 \\ 0 & 1 & 0 \\ -4 & -8 & 1 \end{bmatrix} \qquad \boldsymbol{T}^{-1} = \begin{bmatrix} 1 & 0 & 0 \\ 0 & 1 & 0 \\ 4 & 8 & 1 \end{bmatrix}$$

$$\boldsymbol{R} = \begin{bmatrix} \cos 60° & \sin 60° & 0 \\ -\sin 60° & \cos 60° & 0 \\ 0 & 0 & 1 \end{bmatrix} = \begin{bmatrix} 1 & 0 & 0 \\ -1 & 0 & 0 \\ 0 & 0 & 1 \end{bmatrix}$$

$$\boldsymbol{TRT}^{-1} = \begin{bmatrix} 1 & 0 & 0 \\ 0 & 1 & 0 \\ -4 & -8 & 1 \end{bmatrix}$$

The transformed vertices are (9, 2), (2, 9) and (11, 12).

Exercise 8.4

1. (a) $y-3x+5=0$ (b) $y-3x+4=0$
2. $y=4x$
3. The equation of the line passing through the points (2, 7) and (−4, −5) is $y-2x-3=0$.
   Where the line cuts the x axis the y co-ordinate is 0, hence we have $y-3=0$ so $y=3$. So the line cuts the y axis at (0, 3).

4. The equation $3y + 9x - 6 = 0$ may be written as $y = -3x + 2$. Hence we see that any line parallel to this one will have a gradient of -3. So the equation of such a line will be of the form $y = -3x + c$. The line cuts the y axis at (0, 5) hence $c = 5$. So the equation of this line is $y = -3x + 5$.

Exercise 8.5

1. $x^2 + y^2 = 16$
2. Equation of the circle is $x^2 - 6x + y^2 - 8y = 24$
   Completing the squares gives $(x - 3)^2 - 9 + (y - 4)^2 - 16 = 24$ which gives $(x - 3)^2 + (y - 4)^2 = 49$. The centre of the circle with this equation is at (3, 4). The distance between (3, 4) and (0, 0) is 5.
3. Transformation matrix is:

$$\begin{bmatrix} \cos 45° & \sin 45° & 0 \\ -\sin 45° & \cos 45° & 0 \\ 0 & 0 & 1 \end{bmatrix} \text{ which is equivalent to } \begin{bmatrix} \sqrt{2}/2 & \sqrt{2}/2 & 0 \\ -\sqrt{2}/2 & \sqrt{2}/2 & 0 \\ 0 & 0 & 1 \end{bmatrix}$$

   Centre at (4, 4) is transformed to $(0, 4\sqrt{2})$. The equation of the transformed circle is $x^2 + (y - 4\sqrt{2})^2 = 4$.

Self-Assessment Questions

8.1 1. Since the point P and the origin O* are the same location the co-ordinate of P relative to O* must be (0).
2. The translation equation is $x_P^* = x_P + T_P$. Since we want $x_P^* = -x_P$ we get $-x_P = x_P + T_P$. Hence the translation $T_P = -2x_P$.

8.2 Distance $= \sqrt{x_P^2 + y_P^2}$

8.3

1. $\begin{bmatrix} 1 & 0 & 0 \\ 0 & 1 & 0 \\ -x_P & -y_P & 1 \end{bmatrix}$ 2. $\begin{bmatrix} 1 & 0 & 0 \\ 0 & 1 & 0 \\ 0 & 0 & 1 \end{bmatrix}$

3. No. It is only true when the translation and rotation are such that the matrices $\boldsymbol{T}$ and $\boldsymbol{R}$ representing these transformations satisfy the equation $\boldsymbol{TR} = \boldsymbol{RT}$. As you know from Chapter 7 the operation of multiplication on matrices is not generally commutative and so the equation is not generally true.

8.4 1. $y = b$. 2. The gradients of parallel lines are equal. 3. The gradient of the line is -1 and its equation is $y = -x + a$.

8.5 The circle cuts the x axis at $(r, 0)$ and $(-r, 0)$ and the y axis at $(0, r)$ and $(0, -r)$. Joining these points would generate a square.

Solutions to Chapter 9

Exercise 9.1

1. When $x = 3$, $y = -14$, and when $x = 3.5$, $y = -17$. Thus the change in y is $-17 - (-14) = -3$ and the change in x is 0.5, so the average rate of change of y with respect to x is -6.

2. When $t = 24$, $s = 6$, and when $t = 52$, $s = 8$. The average rate of change of s with respect to t is therefore $2/28 = 1/14$.
3. When $v = c$, $w = c^2 + 2c$. When $v = c + h$, $w = c^2 + 2ch + h^2 + 2c + 2h$, so the average rate of change of w with respect to v is $2c + 2 + h$.
4. When $t = 2$, $n = 56$, and when $t = 4$, $n = 64$. The average rate of change of n with respect to t is therefore $(64 - 56)/2 = 4$, i.e. on average four people arrive each week over that period.

Exercise 9.2

1. (i) 2 (ii) $8x$ (iii) $6x^2$
2. When the radius is $r + \delta r$, the area is $\pi(r^2 + 2r\delta r + (\delta r)^2)$. Subtracting πr^2 from this gives $\pi(2r\delta r + (\delta r)^2)$, and dividing this by δr gives $\pi(2r + \delta r)$. Letting δr tend to zero, we obtain $2\pi r$ for the derivative of A with respect to r. When $r = 4$, then, the rate of change of area with respect to radius is 8π.

Exercise 9.3

1. The derivative we need is $8x$, so the approximate change is, in each case, $16\delta x$.

   When $\delta x = 0.1$, the exact change is 1.64, and the approximate change is 1.6, so the percentage error is about 2.43%.

   When $\delta x = 0.01$, the exact change is 0.1604 and the approximate change is 0.16, so the error is about 0.25%.

   When $\delta x = 0.001$, the exact change is 0.016004 and the approximate change is 0.016, so the error is about 0.025%.

   The percentage error is smallest when the change is smallest.
2. The derivative is $6x^2 - 3$. (i) When $x = 0$, the derivative is -3, which is negative. Hence if x is increased slightly, y will decrease. (ii) When $x = 4$, the derivative is 93, which is positive. Hence if x is decreased slightly, y will decrease.

Exercise 9.4

1. (i) $-6x^{-2}$ (ii) $12r$ (iii) $-1.5y^{-1.5}$
2. (a) (i) $4x - 3$ (ii) $2 + 10t$ (iii) $-4x^{-3} - 0.5x^{-0.5}$ (iv) $4 - 0.25t^{-1.5}$

   (b) The derivative is $2x - 3x^2$, so when $x = 2$ the derivative is $4 - 12 = -8$. The derivative is the gradient of the tangent, so at $x = 2$ the tangent has gradient -8.

   (c) The derivative is $2x + 2$, so when $x = -2$ we find that the tangent has slope -2. Since $2y + 4x - 3 = 0$ is equivalent to $y = -2x + 1.5$, this line has the same slope as the tangent when $x = -2$, i.e. the two are parallel.

   (d) The derivative of $400m - 10m^2$ is $400 - 20m$. If $m = 10$, the derivative is 200, which is positive, so sales are increasing. If $m = 30$, the derivative is -200, which is negative, so sales are decreasing.

Exercise 9.5

1. (a) Here the derivative is $3x^2 - 2x = x(3x - 2)$. This is zero if $x = 0$ or $x = 2/3$. For x near zero, the derivative is approximately $-2x$, so as x increases through 0, the derivative is positive, then zero, then negative. Thus there is a local maximum here. When x is near $2/3$, the derivative is approximately $\frac{2}{3}(3x - 2)$, so as x increases through $2/3$ the derivative is negative, then zero, then positive. Thus we have a local minimum. (b) This time the derivative is $-3x^2$. This is zero only when $x = 0$, and is negative for non-zero x. Thus the stationary point is a saddle.
2. The derivative of $64t - 16t^2$ is $64 - 32t$, which is zero only when $t = 2$. If $t < 2$ the derivative is positive, and if $t > 2$ the derivative is positive. Thus the shape of the graph is an inverted bowl with its apex at $t = 2$. This tells us that the overall maximum value of p must be when $t = 2$, as required.
3. (a) The derivative is $4x^3 - 12x^2 = 4x^2(x - 3)$. This is zero when $x = 0$ or $x = 3$ and there is a saddle at $x = 0$ and a local minimum when $x = 3$.
   (b) Here the derivative is $3x^2 - 6x = 3x(x - 2)$ which is zero when $x = 0$ or $x = 2$. There is a local maximum when $x = 0$ and a local minimum when $x = 2$.
   (c) Here the derivative is $3 - 9v^2 = 3(1 - v)(1 + v)$ which is zero when $v = \pm 1$. When $v = -1$ there is a local minimum and when $v = 1$ there is a local maximum.
   (d) The derivative is $1 - t^{-2} = (1 - t^{-1})(1 + t^{-1})$, which is zero when $t = \pm 1$. When $t = -1$ there is a local maximum and when $t = 1$ there is a local minimum.
4. The derivative of H with respect to i is $8 - 2i$, which vanishes when $i = 4$ and has a local maximum there. When $i = 0$, $H = 0$, when $i = 4$, $H = 16$, and when $i = 6$, $H = 12$. Thus H has its maximum value of 16 when $i = 4$.
5. The derivative of C is $1/20 - 180/n^2$, which vanishes when $n = 60$. For smaller values of n the derivative is negative and for larger values it is positive, which tells us that this gives the overall minimum of C for positive n. Thus the system costs least per user when there are 60 users.

Exercise 9.6

1. We know that the derivative of t^3 is $3t^2$, so it follows that $\frac{2}{3}t^3$ is a primitive for $2t^2$. Thus $\int 2t^2 \mathrm{d}t = \frac{2}{3}t^3 + C$.
2. It is obvious that $3t$ is a primitive for v here, so the distance travelled is $[3t]_2^5 = 15 - 6 = 9$ metres. Alternatively, the area is that of a rectangle with base 3 and height 3, so we obtain an area of $3 \times 3 = 9$.
3. $\displaystyle\int_0^2 x + x^2 \mathrm{d}x = \left[\frac{1}{2}x^2 + \frac{1}{3}x^3\right]_0^2 = \frac{14}{3}$
4. These graphs cross at $x = 0$ and $x = 2$, so we need to find the area above $y = x^4$ and below $y = 4x^2$ between $x = 0$ and $x = 2$.

$$\begin{aligned}\int_0^2 4x^2 - x^4 \mathrm{d}x &= \left[\frac{4}{3}x^3 - \frac{1}{5}x^5\right]_0^2 \\ &= \frac{4}{3}8 - \frac{1}{5}32 = \frac{64}{15}\end{aligned}$$

Self-Assessment Questions

9.1 This average rate of change is the gradient of the line joining the points $(0, y(0))$ to $(1, y(1))$.

9.2 For a sphere, the derivative of volume with respect to radius is the surface area. This is not a coincidence; if you imagine a thick skin on a balloon its volume should be the surface area times the thickness. Thus the increase in volume divided by the increase in radius is approximately the surface area.

9.3 The best approximation to the change in the value of the dependent variable is zero – in the sense that any other ratio of change in independent variable to change in dependent variable gives a percentage error that does not approach zero when the change in the independent variable is made small.

9.4 Consider the function given by $y = x \times x$. Then the derivative of y with respect to x is $2x$, but since the derivative of x with respect to x is 1, the product of the derivatives is 1.

9.5 The second derivative tells us whether the first derivative is increasing or decreasing. For example, if the first derivative is zero and the second is negative, then for slightly smaller values of the independent variable the first derivative must be positive, and for slightly larger values it must be negative; hence there is a local maximum. Similarly, if the second derivative is positive when the first is zero, there must be a local minimum.

9.6 The area comes out as -12.5, so that areas below the x axis are negative. This is sensible because we want $\int_a^b (f(x) - f(x))\mathrm{d}x$ to be zero for any function f.

Index